EXPERIMENTAL TECHNIQUES
IN
FRACTURE MECHANICS

SOCIETY FOR EXPERIMENTAL STRESS ANALYSIS MONOGRAPH SERIES

EXPERIMENTAL TECHNIQUES
IN
FRACTURE MECHANICS

Edited by
ALBERT S. KOBAYASHI
University of Washington
Department of Mechanical Engineering
Seattle, Washington

———————————————————— Published Jointly by ————
THE IOWA STATE UNIVERSITY PRESS
Ames, Iowa
SOCIETY FOR EXPERIMENTAL STRESS ANALYSIS
Westport, Connecticut

SESA Monograph No. 1
This monograph is published in furtherance of SESA objectives in the field of
experimental mechanics. The Society is not responsible for any statements made
or opinions expressed in its publications.

Library of Congress Cataloging in Publication Data

Kobayashi, Albert S   1924–
    Experimental techniques in fracture mechanics.

    (SESA monograph no. 1)
    Includes bibliographical references.
    1. Fracture of solids. 2. Materials—Testing.
I. Title. II. Series: Society for Experimental Stress Analysis. SESA monograph
no. 1. TA409.K62   620.1′126   72-13967   ISBN 0-8138-0710-7

First edition, 1973

CONTENTS

PREFACE

One of the primary objectives of the Society for Experimental Stress Analysis has been the dissemination of information in the field of experimental mechanics. In the past, this has been accomplished through direct exchanges at the Society's semiannual meetings or through its publications, such as EXPERIMENTAL MECHANICS, the SESA Proceedings, and the Handbook of Experimental Stress Analysis. The first-named publication tends to emphasize research papers on advanced topics or applications papers on very specialized subjects. Although excellent textbooks are available which treat some fields of experimental mechanics in a systematic way, gaps still exist and many important subjects are not covered by a uniform treatment in the depth and detail that they deserve.

Several separate series of monographs are currently being planned which will eventually cover the entire field of experimental mechanics. Within each series, the first monograph will begin at a level which can easily be comprehended by an engineer who is not a specialist in the particular area treated. Subsequent individual monographs will proceed at increasing levels of sophistication, until the last monograph in each series may represent the current state of the art. The arrangement of the subject matter into several short but coordinated monographs offers two major advantages. It will allow the individual already in the field to purchase a document on a specialized subject without the necessity of purchasing the entire series. Second, it will be possible to update monographs in rapidly developing fields without the great task and expense (to the Society and the individual) of revising the entire series, much of which may not have changed significantly. In addition to the present monograph, series of monographs on strain gages and photoelasticity are well underway. Monographs on holography and a sequel on fracture mechanics are currently in the planning stages.

The present monograph, *Experimental Techniques in Fracture Mechanics*, is the first to reach the publication stage. The committee is deeply indebted to Professor Albert S. Kobayashi who has carried almost the entire burden of planning and executing the project and, in addition, has contributed three chapters. In spite of careful planning and close coordination in the writing stages, it was still necessary to ask the authors to prepare several drafts of the individual chapters in order to attain the degree of continuity and compatibility desired. The committee is extremely grateful to all the authors for their patience and complete cooperation.

Because of the importance of establishing the series, the committee was particularly anxious that this monograph be as complete and authoritative as possible. Consequently, because of his eminence in this field, Dr. George R. Irwin was asked to review all manuscripts and to make comments. He graciously accepted that request and many of his suggestions were included by the authors in the final versions of their chapters.

Finally, the committee wishes to express its gratitude to Mr. B. E. Rossi, managing director of the Society for Experimental Stress Analysis, for his encouragement and cooperation. Mr. Rossi also accepted the responsibility for the editorial styling and for working with the publishers on the many time-consuming details that are an essential part of any endeavor such as this.

The SESA Committee on Monographs
C. E. Taylor, Chairman
P. H. Adams
J. W. Dally
A. S. Kobayashi
M. M. Lemcoe
M. M. Leven
W. F. Riley
L. J. Weymouth

1
INTRODUCTION

by

Albert S. Kobayashi
University of Washington
Dept. of Mechanical Engineering
Seattle, Wash.

Although spectacular brittle fracture of structural components has occurred during many centuries, an often-quoted classical example is the failure of a molasses tank at noon on January 15, 1919, in Boston, Mass.[1] The 2,300,000 gallons released from this tank killed 12 persons, injured 40, and drowned a number of horses. Prominent scientists of the day investigated this incident with inconclusive results. Since that time, serious fracture problems have occurred with airplanes, large rotors, armored vehicles, guns, storage tanks, pressure vessels, pipelines, welded ships,[2] and aerospace structures.[3]

The catastrophic possibilities of sudden fracture in terms of financial loss, human life, and even national prestige and security have inspired many studies on fracture phenomena. Among the disciplines related to fracture studies, the group of techniques which are termed fracture mechanics has been developed to a point where it is used in advanced design of hardware as well as in fracture-failure investigations. Readers are referred to a recent treatise on fracture that encompasses the entire field in seven volumes for a comprehensive study of fracture mechanics.[4]

In addition to the references quoted in this chapter, up-to-date information is published in the following two journals which deal primarily with topics in fracture mechanics: *International Journal of Fracture Mechanics*, Walters-Noordhoff; and *International Journal of Engineering Fracture Mechanics*, Pergamon Press.

In addition, technical papers on fracture mechanics appear frequently in *Experimental Mechanics*, SESA; *Journal of Basic Engineering, Transactions of ASME*, ASME; *Materials Research and Standards*, ASTM; and *Journal of Materials*, ASTM.

Technical and professional societies throughout the world have committees and task forces organized for the specific purpose of studying problems related to fracture. Among the most active ones is the Committee E-24 on Fracture Testing of Metals under the American Society for Testing and Materials. This committee is subdivided into six subcommittees which deal with topics such as test methods in fracture mechanics, fractography and microstructure, dynamics test methods, subcritical crack growth, nomenclature and definitions, and applications.

The purpose of this SESA monograph is to assemble, in several chapters, experimental techniques related to fracture mechanics which are scattered throughout the many publications and committee reports stated above. The material contained in this first monograph is intended to provide information on the principles, procedures, applications and, finally, limitations on some basic experimental techniques currently used in experimental fracture mechanics. Standard experimental details such as fracture-specimen configurations and loading procedures are amply described in special publications of the American Society for Testing and Materials[5-9] and are, thus, not elaborated in this monograph. Rather, it is hoped that by limiting the coverage of each topic, the reader can obtain an overview of representative experimental techniques in fracture mechanics, and then turn to the quoted references for specific details which fit his particular requirement. Thus, this monograph is primarily written for the initiates in fracture mechanics, but it also contains advanced and current information for the experts in this field.

Five major areas are covered in this monograph. These are (1) theory of fracture mechanics; (2) acoustic-emission techniques; (3) compliance measurements; (4) testing systems and associated instrumentation; and (5) photoelastic techniques. By following these chapters in sequence, the reader is provided with some theoretical background in fracture mechanics in Chapter 2, learns an experimental technique which shows great potential in detecting onset of fracture in Chapter 3, studies an experimental technique for determining the strain-energy release rate in Chapter 4, is indoctrinated with some of the unexpected problems in designing a test system for fracture testing in Chapter 5, and is finally introduced to photoelastic techniques for determining stress-intensity factors in Chapter 6.

Finite-element analysis has also been included in Chapter 2 as an experimental tool in the broad sense which also reflects SESA's recently expanded role to include topics in finite-element analysis. The mutually complimentary roles of such numerical techniques with experimental techniques are emphasized in this monograph.

In assembling this monograph, this editor was very fortunate in having contributions from internationally known experts in their respective fields. The monograph was reviewed by Professor Charles E. Taylor, chairman of the SESA Monograph Committee, and by Professor James W. Dally, 1970–1971 president of SESA, both of whom suggested changes which enhanced this monograph and for which the authors are very grateful.

Finally, the authors wish to acknowledge the many constructive comments and suggested revisions by Professor George R. Irwin, University of Maryland, who also reviewed this monograph in detail.

The task of the contributors and of the editor has not terminated with the publication of this first monograph. The content of this monograph will be updated periodically in order to maintain it current at all times.

REFERENCES

1. Shank, M. E., "A Critical Survey of Brittle Fracture in Carbon Plate Steel Structure Other Than Ships," published by Welding Research Council of the Engineering Foundation, New York (1954). See also ASTM STP **158**, 45 (1954).
2. Liebowitz, H., ed., *Fracture, an Advanced Treatise*, **5**, Design of Structures, Academic Press (1969).
3. Tiffany, C. F., and Masters, J. N., *Applied Fracture Mechanics*, ASTM STP **381**, 249–278 (1965).
4. Liebowitz, H., ed., *Fracture, an Advanced Treatise*, **1–7**, Academic Press (1969, 1970).
5. *Fracture Toughness Testing and Its Application*, ASTM STP 381 (1965).
6. *Plane Strain Crack Toughness Testing at High Strength Metallic Materials*, ASTM STP **410** (1966).
7. *Fatigue Crack Propagation*, ASTM STP 415 (1967).
8. Brown, W. F., Jr., ed., *Review of Developments in Plane Strain Fracture Toughness Testing*, ASTM STP **463** (1970).
9. *Damage Tolerance in Aircraft Structure*, ASTM STP 486 (1971).

2

FRACTURE MECHANICS

by

Albert S. Kobayashi
University of Washington
Dept. of Mechanical Engineering
Seattle, Wash.

2.1 INTRODUCTION

Structural failures under loading conditions well below the yield stress of the structural material can often be attributed to cracks or cracklike flaws in the structure. Such failures show that the conventional strength analysis of structures alone, no matter how accurately conducted, is not sufficient to guarantee the structural integrity under operational conditions. Structural study which considers crack-extension behavior as a function of applied loads is called fracture mechanics. In particular, in the absence of large plastically yielded regions surrounding cracks or flaws, such study is referred to as linear fracture mechanics.

As a result of considerable research efforts during the past decade, linear fracture mechanics can now be used to solve many practical engineering problems in failure analysis, material selection, structural-life prediction, and acceptance tests. Linear fracture mechanics can also be extended to solve fracture problems involving moderate plastic yielding by incorporating various plasticity correction factors provided fracture occurs prior to large-scale yielding of the structural member. An excellent historical review of the development of fracture mechanics starting with Ludwik's paper in 1900 through Inglis, Griffith, Stanton and Batson, Doherty, Westergaard, Sneddon, Sach, Fisher and Holloman, Irwin, Orowan, Irwin and Kies, Wells, and Williams's paper of 1957 is given in Ref. 1.

The basis of linear fracture mechanics is the paper of Griffith[2] which was unrecognized for thirty years. The original Griffith theory as proposed in 1921 stated that the rupture of brittle material, i.e., glass, occurred when the surface area, A, of a flaw in a body under load as shown in Fig. 2.1 enlarged to an area of $A + \delta A$ with

$$|\delta U_\sigma| > |\delta U_{st}| \tag{2.1}$$

4

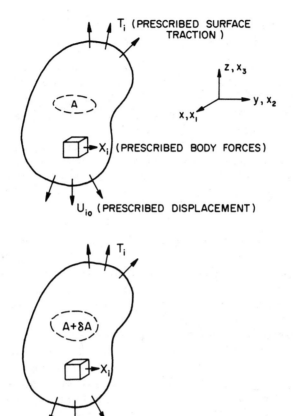

Fig. 2.1—Elastic body with enlarging crack.

where δU_σ = the decrease in potential energy due to increased crack-surface area of δA

δU_{st} = the increase in surface energy due to increased crack-surface area.

The terms above involving increase in energy quantities are evaluated with respect to the increased crack-surface area δA and, therefore, the δ quantities are referred to as rate from here on.

Irwin[3] and Orowan[4] subsequently modified the original Griffith theory so that it could be applied to metals by adding a term involving the plastic-energy-dissipation rate in the plastic region surrounding the crack tip. The change of total energy balance involving input-energy change and the dissipated-energy change due to increase of crack-surface area, δA, was expressed as

$$\delta W + \delta U_\sigma \geq \delta U_{st} + \delta P \tag{2.2}$$

where δW = the increased rate of external work

   δP = the increased rate in plastic-energy dissipation.

Irwin recognized that the plastic-energy-dissipation rate is much larger than the surface-energy-dissipation rate and, therefore, proposed to ignore the latter. The Griffith-Irwin theory of fracture then states that, when the input-energy rate, $\delta W + \delta U_\sigma$, exceeds the dissipated plastic-energy rate, δP, instability occurs and the crack will run. For a constant-displacement boundary condition which is commonly referred to as fixed-grip loading condition in fracture testing, no additional work is done to the system or $\delta W = 0$. Thus, the stability criterion represented by eq. (2.2) becomes

$$\delta U_\sigma \geq \delta P \qquad (2.3)$$

It can be easily proven at this point that the input-energy rate for an infinitesimal crack extension is independent of the load application whether it is a fixed-grip condition, a fixed-force condition, or a combination of these two.[4] This input-energy rate is, thus, commonly referred to as the strain-energy-release rate, $\mathcal{G}$, for a unit area increase in the cracked surface of the cracked body.

Irwin also postulated that the plastic-dissipation rate, δP, is a material property which can be determined by standardized tests at the onset of fracture or when the equality sign holds in eq. (2.3). Since the left-hand side of eq. (2.3), δU_σ, depends on the loading condition and crack geometry, the problem in linear fracture mechanics now reduces to a boundary-value problem in determining δU_σ for various crack problems.

The above theory by Griffith on energy balance and its subsequent modifications by Irwin and Orowan are a necessary condition for the onset of fracture.[5] As such, Griffith's theory cannot conveniently characterize all types of fracture observed in fracture tests. Irwin thus proposed that the local stress field surrounding the crack tip be used in place of the total input-energy rate for such characterization.[6]

The relation between δU_σ mentioned above and the local stress field can be obtained by considering the reverse problem where a short segment δx of a two-dimensional crack is closed by imposing a force of $\sigma_{yy}^*(-\delta x, 0)$ on the crack surface as shown in Fig. 2.2. The total strain-energy absorption rate in this reverse-loading problem is equal to the input-energy rate, δU_σ, which, in turn, is equal to the strain-energy-release rate, $\mathcal{G}$, for a crack extension of δx or

$$\delta U_\sigma = \mathcal{G} \cdot \delta x = \int_0^{\delta x} u_y(0,0) \cdot \sigma_{yy}^*(-\delta x, 0) \, dx \qquad (2.4)$$

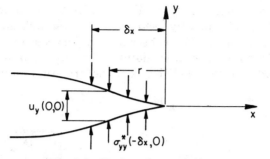

Fig. 2.2—Closing of a crack tip.

where $u_y(0,0)$ is the crack-closing displacement when the crack closes from the origin of the x-y coordinate for a length of δx.

Assuming further that the plastically yielded region ahead of the crack tip does not change the state of stress significantly, then $u_y(0,0)$ and $\sigma_{yy}^*(-\delta x,0)$ can be determined by the elastic displacements and elastic stresses obtained through the known elastic state in the vicinity of a crack tip. In particular, $u_y(0,0)$ is the crack-closing displacement with the crack tip located at the origin of the coordinates, $(0,0)$, and $\sigma_{yy}^*(-\delta x,0)$ is the associated closing stress which is equal to the normal stress ahead of the crack tip located at $(-\delta x,0)$. Thus, all problems in linear fracture mechanics can now be converted to problems in linear elasticity.

In the following sections, it will be shown how some relations in fracture mechanics can be derived by the use of linear theory of elasticity. The purpose here is to prepare the readers for the main sections of the monograph by presenting a brief sketch of some useful theoretical results in linear fracture mechanics. For more complete coverage, readers are referred to Refs. 6 through 9.

2.2 TWO-DIMENSIONAL PROBLEMS

Westergaard's Stress Functions

Airy's stress function which is a biharmonic stress function used in solving two-dimensional boundary-value problems can be represented by complex variables as

$$\Phi_I = Re\,\overline{\overline{Z}}_I + y\,Im\,\overline{Z}_I \qquad (2.5)$$

where $\overline{Z}_I = \dfrac{d\overline{\overline{Z}}_I}{dz}$ and $Z_I = \dfrac{d\overline{Z}_I}{dz}$

$z = x + iy$

The above complex function, $Z_I(z)$, referred to as Westergaard's stress function, is frequently used to solve two-dimensional problems in cracked structures. The subscript I in Airy's stress function and Westergaard's stress function defined by eqs. 2.5 represents the "opening mode" or "mode I" crack-tip deformation shown in Fig. 2.3.

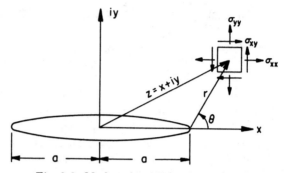

Fig. 2.3—Modes of crack-tip deformation.

The associated stress components are then

$$\sigma_{xx} = Re\ Z_I - y\ Im\ Z_I'$$
$$\sigma_{yy} = Re\ Z_I + y\ Im\ Z_I'$$
$$\sigma_{xy} = -y\ Re\ Z_I' \tag{2.6}$$

where $Z_I' = \dfrac{dZ_I}{dz}$

It can also be shown that Westergaard's stress function for the mode I crack-tip deformation with crack-tip locations at $(-a, o)$ and (a, o) as shown in Fig. 2.4 can be represented as

$$Z_I = \frac{g(z)}{\sqrt{z^2 - a^2}} \tag{2.7}$$

where $g(z)$ is an analytic function dependent on the outer geometry

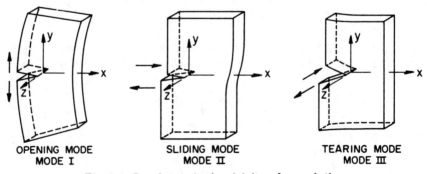

OPENING MODE   SLIDING MODE   TEARING MODE
MODE I   MODE II   MODE III

Fig. 2.4—Local state in the vicinity of a crack tip.

of the plane and the associated boundary conditions. For a uniformly applied stress of σ on an infinite plane, it can be easily shown that $g(a) = \sigma a$.

If we further confine ourselves to a small region in the vicinity of a crack tip also shown in Fig. 2.4 and represent all stresses and stress functions in terms of a local polar-coordinate system of r and θ, Westergaard's stress function becomes

$$Z_I = \frac{f(\rho)}{\sqrt{\rho}} \qquad (2.8)$$

where $\rho = re^{i\theta}$

As $\rho \rightarrow 0$, however, the complex analytic function, $f(\rho)$, for the opening mode approaches a real constant, $K/\sqrt{2\pi}$, where K is the opening mode of stress-intensity factor and is defined as

$$K = \lim_{r \to 0} \sigma_{yy}(r, \theta = 0) \cdot \sqrt{2\pi r} \qquad (2.9)$$

The local stresses in terms of local coordinates are[9]

$$\sigma_{xx} = \frac{K}{\sqrt{2\pi r}} \cos \frac{\theta}{2} \left[1 - \sin \frac{\theta}{2} \cdot \sin \frac{3\theta}{2} \right]$$

$$\sigma_{yy} = \frac{K}{\sqrt{2\pi r}} \cos \frac{\theta}{2} \left[1 + \sin \frac{\theta}{2} \cdot \sin \frac{3\theta}{2} \right]$$

$$\sigma_{xy} = \frac{K}{\sqrt{2\pi r}} \sin \frac{\theta}{2} \cdot \cos \frac{\theta}{2} \cdot \cos \frac{3\theta}{2} \qquad (2.10)$$

For the state of plane stress, the displacement components in the x and y directions, u_x and u_y, become

$$u_x = \frac{K}{G} \sqrt{\frac{r}{2\pi}} \cos \frac{\theta}{2} \left[\frac{1 - \nu}{1 + \nu} + \sin^2 \frac{\theta}{2} \right]$$

$$u_y = \frac{K}{G} \sqrt{\frac{r}{2\pi}} \sin \frac{\theta}{2} \left[\frac{2}{1 + \nu} - \cos^2 \frac{\theta}{2} \right] \qquad (2.11a)$$

where G = the shear modulus
ν = the Poisson's ratio

For the state of plane strain, the displacement components become

$$u_x = \frac{K_I}{G} \sqrt{\frac{r}{2\pi}} \cos \frac{\theta}{2} \left[1 - 2\nu + \sin^2 \frac{\theta}{2} \right]$$

$$u_y = \frac{K_I}{G} \sqrt{\frac{r}{2\pi}} \sin \frac{\theta}{2} \left[2 - 2\nu - \cos^2 \frac{\theta}{2} \right] \qquad (2.11b)$$

The added subscript of I onto the opening mode of stress-intensity factor, K, denotes the state of plane strain when discussion is restricted to tensile fracture. Likewise, designation does not exist for sliding mode or mode II crack-tip deformation* where the plane-stress and plane-strain stress-intensity factors are both referred to as K_{II}.

Having established the states of displacements and stresses in the vicinity of the crack tip, the strain-energy-release rate due to crack extension can be determined by knowing $u_y(0,0)$ and $\sigma_{yy}^*(-\delta x,0)$ evaluated at r distance from the crack tip as

$$\sigma_{yy}^*(-\delta x,0) = \frac{K_I}{\sqrt{2\pi(\delta x - r)}}$$

$$u_y(0,0) = \frac{2(1-\nu)}{G} \frac{K_I}{\sqrt{2\pi}} \sqrt{r} \qquad \text{(Plane strain)} \qquad (2.12)$$

Substitution of eq. (2.12) into eq. (2.4) yields the strain-energy-release rate due to a crack extension of δx in a state of plane strain as

$$\delta U_\sigma = \mathcal{G}_I \delta x = \frac{1-\nu}{G} K_I^2 \int_0^{\delta x} \sqrt{\frac{r}{\delta x - r}}\, dr = \frac{1-\nu}{2G} K_I^2\, \delta x$$

where $\mathcal{G}_I$ represents the strain-energy-release rate under plane-strain conditions. The characteristic quantity of strain-energy-release rate is also referred to as a crack-opening force for the state of plane strain and can be represented as

$$\mathcal{G}_I = \frac{1-\nu}{2G} K_I^2 \qquad (2.13)$$

The strain-energy-release rate for the "state of plane stress" can be obtained by replacing ν with $\nu/(1+\nu)$ in the above equation which yields $\mathcal{G} = K^2/E$ where E is the modulus of elasticity.

The plastic-energy-dissipation rate which is a material constant following Griffith-Irwin theory can now be replaced by an equivalent material constant which is the critical strain-energy-release rate at the onset of rapid fracture or $\mathcal{G}_{IC}$. Since the stress-intensity factor, K_I, is directly related to $\mathcal{G}_I$ from eq. (2.13), the Griffith-Irwin theory can be restated in terms of the stress-intensity factor K_I which becomes a material constant, K_{Ic}, at the onset of rapid fracture. This critical stress-intensity factor, K_{Ic}, is referred to as the fracture toughness of the material and has the dimension of stress $\cdot \sqrt{\text{length}}$.

---

*See Fig. 2.3.

For the "forward-sliding mode" of the "mode II" of crack deformation, Westergaard's stress function, Z_{II}, and the stress components as well as displacement components can be represented by equations comparable to eqs. (2.6) through (2.11). For example, Airy's stress function for the sliding mode takes the form of[8]

$$\phi_{II} = -y \, Re \, \overline{Z}_{II} \qquad (2.14)$$

For a limiting smallness of ρ, the stress and displacement components in terms of the local polar-coordinate systems shown in Fig. 2.4 become

$$\sigma_{xx} = \frac{-K_{II}}{\sqrt{2\pi r}} \sin \frac{\theta}{2} \left[2 + \cos \frac{\theta}{2} \cdot \cos \frac{3\theta}{2} \right]$$

$$\sigma_{yy} = \frac{K_{II}}{\sqrt{2\pi r}} \cos \frac{\theta}{2} \cdot \sin \frac{\theta}{2} \cdot \cos \frac{3\theta}{2}$$

$$\sigma_{xy} = \frac{K_{II}}{\sqrt{2\pi r}} \cos \frac{\theta}{2} \cdot \left[1 - \sin \frac{\theta}{2} \sin \frac{3\theta}{2} \right] \qquad (2.15)$$

For the state of plane stress, the displacement components become

$$u_x = \frac{K_{II}}{G} \sqrt{\frac{r}{2\pi}} \sin \frac{\theta}{2} \left[\frac{2}{1+\nu} + \cos^2 \frac{\theta}{2} \right]$$

$$u_y = \frac{K_{II}}{G} \sqrt{\frac{r}{2\pi}} \cos \frac{\theta}{2} \left[\frac{1-\nu}{1+\nu} + \sin^2 \frac{\theta}{2} \right] \qquad (2.16a)$$

For the state of plane strain

$$u_x = \frac{K_{II}}{G} \sqrt{\frac{r}{2\pi}} \sin \frac{\theta}{2} \left[2 - 2\nu + \cos^2 \frac{\theta}{2} \right]$$

$$u_y = \frac{K_{II}}{G} \sqrt{\frac{r}{2\pi}} \cos \frac{\theta}{2} \left[1 - 2\nu + \sin^2 \frac{\theta}{2} \right] \qquad (2.16b)$$

where the stress-intensity factor for mode II crack-tip displacement is also obtained by

$$K_{II} = \lim_{r \to 0} \sigma_{xy}(r, \theta = 0) \cdot \sqrt{2\pi r} \qquad (2.17)$$

The strain-energy-release rate for the mode II deformation takes on an identical form as eq. (2.13) as

$$\mathcal{G}_{II} = \frac{1-\nu}{2G} K_{II}^2 \qquad (2.18)$$

In addition to the above two modes of crack-tip deformation, there exists a "parallel sliding" or a "tearing mode," i.e., mode III deformation as shown in Fig. 2.3. This deformation mode is due to an out-of-plane shear which can be produced locally in torsion loading. For such crack-tip deformations, it is convenient to replace Airy's stress function with an out-of-plane warping function $w(x,y)$ which is a harmonic function satisfying the stress equation of equilibrium. By choosing the displacement function as $\overline{Z}_{III}$

$$w = \frac{1}{G} \, Im \, \overline{Z}_{III} \quad \text{and} \quad u_x = u_y = 0 \tag{2.19}$$

Again, for a small ρ, the stresses and displacements in terms of local coordinates become[8]

$$\sigma_{xz} = \frac{K_{III}}{\sqrt{2\pi r}} \sin \frac{\theta}{2} \qquad \sigma_{xx} = \sigma_{yy} = \sigma_{xy} = 0$$

$$\sigma_{yz} = \frac{K_{III}}{\sqrt{2\pi r}} \cos \frac{\theta}{2} \tag{2.20}$$

and

$$u_z = w = \frac{K_{III}}{G} \sqrt{\frac{2r}{\pi}} \sin \frac{\theta}{2} \tag{2.21}$$

where the stress-intensity factor, K_{III}, can be determined by the now familiar form of

$$K_{III} = \lim_{r \to 0} \sigma_{yz}(r, \theta = 0) \cdot \sqrt{2\pi r} \tag{2.22}$$

In comparing eqs. (2.5), (2.14), and (2.19), one notes that the same Westergaard's stress function, Z, can be used to solve geometrically similar problems for the three modes of crack-tip deformation. Finally, one should also note that the theory of fracture mechanics described so far is based on the linear theory of elasticity and, thus, the stress functions for other boundary-value problems with identical geometries and same opening mode can be superimposed. The corresponding stress-intensity factors for the same mode of crack deformation can, therefore, be superimposed as

$$K_I = K_{I1} + K_{I2} + \cdots + K_{Im}$$
$$K_{II} = K_{II1} + K_{II2} + \cdots + K_{IIn}$$
$$K_{III} = K_{III1} + K_{III2} + \cdots + K_{IIIp} \tag{2.23}$$

where m, n, and p are integers.

On the other hand, the crack-opening force is a scalar term involving the strain-energy-release rate and, therefore, crack-opening forces for different modes of crack deformation can be superposed as

$$\mathcal{G} = \mathcal{G}_I + \mathcal{G}_{II} + \mathcal{G}_{III} = \frac{1-\nu}{2G} K_I^2 + \frac{1-\nu}{2G} K_{II}^2$$

$$+ \frac{1}{2G} K_{III}^2 \qquad \text{for plane strain.} \qquad (2.24)$$

Muskhelishvili's Stress Functions

The Muskhelishvili's stress function[10] which is suitable for solving boundary-value problems by conformal-mapping technique has been used extensively in determining stress-intensity factors for various two-dimensional problems. Muskhelishvili's stress function consists of two analytical functions, $\phi(z)$ and $\chi(z)$, which are related to Airy's stress function, $\Phi(x,y)$, through

$$\Phi(x,y) = Re\ [\bar{z} \cdot \phi(z) + \chi(z)] \qquad (2.25)$$

where $\bar{z}$ denotes the complex conjugate of z. The state of stresses and displacements for plane strain can then be expressed as

$$\sigma_{xx} + \sigma_{yy} = 4\ Re\ \{\phi'(z)\}$$

$$\sigma_{xx} - \sigma_{yy} + i2\ \sigma_{xy} = 2[\bar{z}\ \phi''(z) + \chi''(z)]$$

$$u_x + iu_y = \frac{3-4\nu}{2G} \phi(z) - \frac{1}{2G}\ [z\ \overline{\phi'}(\bar{z}) + \overline{\chi'}(\bar{z})] \qquad (2.26)$$

These equations can be reduced to the near-field solutions represented by eqs. (2.10) and (2.11), and (2.15) and (2.16), with the use of the polar-coordinate system shown in Fig. 2.4 and proper approximations.

If one is interested only in the stress-intensity factors, the entire states of stress and displacement need not be determined since the complex stress-intensity factor can be represented as

$$K_I - iK_{II} = 2\sqrt{2\pi} \lim_{\rho \to 0} \sqrt{\rho}\ \phi'(a + \rho) \qquad (2.27)$$

where a is the location of the crack tip on the x-axis as shown in Fig. 2.4. Thus, only the first function of $\phi(z)$ is necessary for K_I and K_{II} determination.

A mapping function which maps a straight crack of length $2a$ onto a unit circle in the ζ-η plane can be represented as

$$z = \omega(\zeta) = \frac{a}{2} \left(\zeta + \frac{1}{\zeta} \right) \qquad (2.28)$$

where $\zeta = \xi + i\eta$

From eq. (2.28), one notes that the crack-tip location $z = \pm a$ is now mapped into $\zeta = \pm 1$. Equation (2.27) which is used to determine the stress-intensity factor then becomes

$$K_I - iK_{II} = 2 \sqrt{\frac{\pi}{a}} \, \phi'(\zeta) \bigg]_{\zeta=1} \tag{2.29}$$

For example, the solution of an infinite plate with a straight crack, and subjected to uniform tension of σ, can be represented by Muskhelishvili's stress functions of which the first part is

$$\phi(z) = \sigma \sqrt{z^2 - a^2} = \frac{\sigma a}{2} (\zeta - 1) \tag{2.30}$$

and the stress-intensity factor becomes

$$K_I - iK_{II} = \sqrt{\pi} \, \sigma \sqrt{a} \tag{2.31}$$

For the case of an infinite plate subjected to uniform shear of τ, the first part of Muskhelishvili's stress function is

$$\phi(z) = i\tau \sqrt{z^2 - a^2} = i \frac{\tau a}{2} (\zeta - 1) \tag{2.32}$$

and the stress-intensity factor becomes

$$K_I - iK_{II} = -i\sqrt{\pi} \, \tau \sqrt{a} \tag{2.33}$$

Finite-element Method

As described in the preceding section, stress-analysis information needed in applying fracture mechanics to actual engineering problems consists of the stress-intensity factors of actual and possible cracks. Unfortunately, the analytical methods cited above fail to yield stress-intensity factors of most cracked surfaces except for a limited number of simple problems. For most industrial applications of fracture mechanics where a 5–10-percent error is tolerated, a numerical solution to actual problems in two dimensions appears to be more palatable than lengthy exact solutions to idealized problems which only approximate the actual conditions. The numerical method which has been widely accepted in industry is the method of finite-element analysis which is described in various textbooks (see, for example, Ref. 11). The application of finite-element analysis to two-dimensional problems in fracture mechanics is, however, new and thus warrants some elaboration.

Theoretical background. The elastic state of stress in the vicinity of the crack tip as expressed in terms of local coordinates can be obtained by the linear superposition of eqs. (2.10) and (2.15) for stresses and eqs. (2.11a) and (2.16a) for displacement in plane stress as

$$\sigma_{xx} = \frac{K}{\sqrt{2\pi r}} \cos \frac{\theta}{2} \left[1 - \sin \frac{\theta}{2} \sin \frac{3\theta}{2} \right]$$

$$- \frac{K_{II}}{\sqrt{2\pi r}} \sin \frac{\theta}{2} \left[2 + \cos \frac{\theta}{2} \cdot \cos \frac{3\theta}{2} \right]$$

$$\sigma_{yy} = \frac{K}{\sqrt{2\pi r}} \cos \frac{\theta}{2} \left[1 + \sin \frac{\theta}{2} \sin \frac{3\theta}{2} \right]$$

$$+ \frac{K_{II}}{\sqrt{2\pi r}} \cos \frac{\theta}{2} \sin \frac{\theta}{2} \cos \frac{3\theta}{2}$$

$$\sigma_{xy} = \frac{K}{\sqrt{2\pi r}} \sin \frac{\theta}{2} \cos \frac{\theta}{2} \cos \frac{3\theta}{2}$$

$$+ \frac{K_{II}}{\sqrt{2\pi r}} \cos \frac{\theta}{2} \left[1 - \sin \frac{\theta}{2} \cdot \sin \frac{3\theta}{2} \right] \tag{2.34}$$

The plane-stress state of displacement in the vicinity of the crack tip can be expressed as

$$u_x = \frac{K}{G} \sqrt{\frac{r}{2\pi}} \cos \frac{\theta}{2} \left[\frac{1 - \nu}{1 + \nu} + \sin^2 \frac{\theta}{2} \right] + \frac{K_{II}}{G} \sqrt{\frac{r}{2\pi}} \sin \frac{\theta}{2}$$

$$\cdot \left[\frac{2}{1 + \nu} + \cos^2 \frac{\theta}{2} \right]$$

$$u_y = \frac{K}{G} \sqrt{\frac{r}{2\pi}} \sin \frac{\theta}{2} \left[\frac{2}{1 + \nu} - \cos^2 \frac{\theta}{2} \right] + \frac{K_{II}}{G} \sqrt{\frac{r}{2\pi}} \cos \frac{\theta}{2}$$

$$\cdot \left[\frac{1 - \nu}{1 + \nu} + \sin^2 \frac{\theta}{2} \right] \tag{2.35}$$

If the states of stress or displacement in the vicinity of the crack tip can be determined within a reasonable degree of accuracy, the stress-intensity factors can be computed by the use of eqs. (2.34) or (2.35). The finite-element analysis must then produce sufficiently accurate states of stress or displacement within the local region where these equations are valid. As a well-known rule of thumb, this local region has been defined as $r < (a/20)$, where a is the half-crack length of a straight crack. Simple calculations of crack-opening displacements (COD) show that this restriction can be relaxed significantly along the crack boundary (or $\theta = \pi$) and that the COD computed by the use of eq. (2.35) would exceed the exact value by 5 percent for $r = a/5$ in a centrally notched, infinite plate subjected to uniaxial tension. This relaxed restriction on the size of local region where eqs. (2.34) or (2.35) are applicable is particularly important in establishing the minimum mesh size for the finite-element analysis.

Recently Watwood[12] and Deverall and Lindsey[13] have shown that the stress-intensity factor can be determined within 1–3-percent accuracy by computing the strain-energy-release rate with crack extension. Basically, the procedure is to compute the total stored strain energy for a given crack geometry and then recompute the total stored strain energy for a small crack extension. The difference between the two total strain energies, ΔU_σ, divided by the difference in surface areas of the two crack lengths, ΔA, will provide the strain-energy-release rate, $\mathcal{G}$, as

$$\mathcal{G} = \lim_{\Delta a \to 0} \frac{\Delta U_\sigma}{\Delta A} \tag{2.36}$$

The advantage of this procedure is that the requirement of minimum grid size at the crack tip can be relaxed without reduction in accuracy. Recent work by this writer has also confirmed this claim.[14]

Numerical procedure. To determine the optimum procedure for evaluating the stress-intensity factor, a finite-width tension plate with a central notch, as shown in Fig. 2.5, was considered. A quadrant of this plate was initially divided into 339 rectangular elements for the coarse-grid analysis. The lines of symmetry where normal displacements and tangential forces vanish are represented by rollers in this figure. Using the results of the coarse-grid analysis, a portion of the plate surrounding the crack tip was analyzed again in a fine-grid analysis, with the prescribed force-boundary conditions established from the coarse-grid analysis. In this fine-grid analysis, 798 elements were used. For each crack length-to-plate width ratio, i.e., a/b, the size of the crowded mesh surrounding the crack tip in Detail A of Fig. 2.5 was determined individually to make the minimum mesh size smaller than $a/10$ for each crack length considered. The computer program used was a standard program in direct-stiffness method available from the University of Washington Computer Center. Typical computer runs for the coarse- and fine-grid analysis on an IBM 7040-7094 system amounted to 3.5 and 6.5 min, respectively. Both approaches by the use of eqs. (2.34) and (2.35) were used to evaluate the stress-intensity factors for this problem.

In using the approach by stresses of eq. (2.34), it was found that the stress-intensity factors were underestimated by about 10–15 percent, primarily due to the inability of the method of finite-element analysis to handle problems with steep stress gradients, such as those

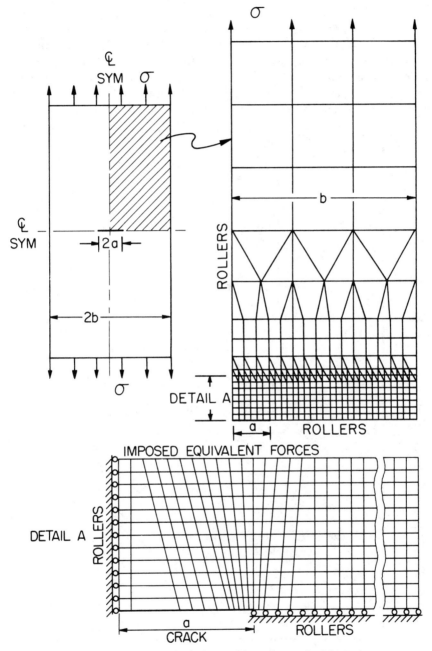

Fig. 2.5—Centrally cracked plate subjected to uniaxial tension.

which exist in the vicinity of a crack tip. Part of this inaccuracy was due to the stiffness matrix used which was derived on the basis of uniform strain and, hence, uniform stress in the basic finite element.

On the other hand, the approach by displacement or eq. (2.35) yielded reasonable results when the largest value of stress-intensity factors computed for several nodes adjacent to the crack tip was used. Figure 2.6 shows a comparison of the stress-intensity factor

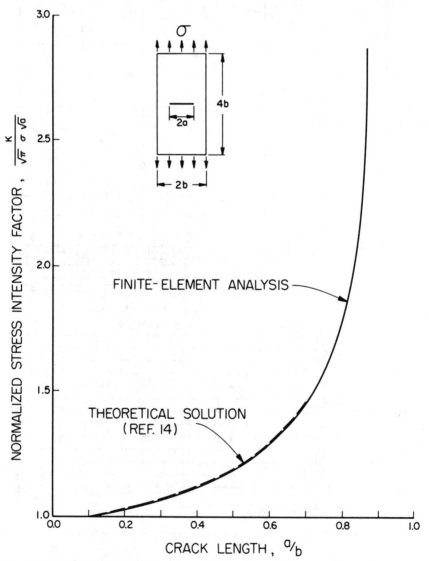

Fig. 2.6—Stress-intensity factor for a finite plate with a central crack and subjected to uniaxial tension.

evaluated by the displacement procedure using the COD with that obtained analytically.[15] In retrospect, the use of COD in place of stress should be a natural approach for the method of direct stiffness which determines the unknown nodal displacements from the known nodal forces through the use of the stiffness matrix.

The theoretical solution in Fig. 2.6 extends only to $a/b = 0.7$ and, therefore, no definite assessment on the accuracy of the present results can be made for $a/b > 0.7$. Similar calculation[14] using the strain-energy-release-rate approach, i.e., eq. (2.36), also yielded results identical to those shown in Fig. 2.6. From our experiences in structural analysis using the finite-element method, there is little reason to believe that the same accuracy cannot be maintained in the region of $a/b > 0.7$.

In the following, two practical problems in two-dimensional fracture mechanics solved by the finite-element analysis are described. In both cases, available analytical techniques could not be used to solve these problems.

Surface-flawed cylinder subjected to internal pressure. A frequently observed flaw in pressure vessels is a surface flaw in the vessel wall.[16] While such flawed wall represents a three-dimensional problem in fracture mechanics, which will be described later, the maximum stress-intensity factor can often be estimated through two-dimensional analysis. The stress-intensity factor of a crack, which penetrates the wall thickness of a pressurized cylinder, has been determined in the past,[17] but no solution exists for the stress-intensity factor of a part-through surface flaw in a pressurized cylinder. For a long surface flaw, the maximum stress-intensity factor at the bottom of the surface flaw can be approximated by its two-dimensional model shown in Fig. 2.7. Figure 2.8 shows the stress-intensity factors in nondimensionalized form for three cases of thickness-to-radius ratio, together with stress-intensity factor for a single-edge notched tension strip.[18] The conflicting effects of circumferential constraint bending, as well as the nonlinear distribution of the hoop stresses, are probably responsible for significant differences between these results and Gross's results.[18]

Slanted crack in plate subjected to uniaxial tension. In fatigue problems, one often encounters cracks which have changed directions due to change in loading conditions. The stress-intensity factors in a slanted crack in a finite-width plate, as shown in Fig. 2.9, can be obtained only by numerical methods. In this problem, two modes of crack-tip deformation exist, i.e., the opening mode and the sliding mode and, therefore, it is necessary to determine K as well as K_{II}. Referring to eq. (2.35), one notes that u_x vanishes for the opening-

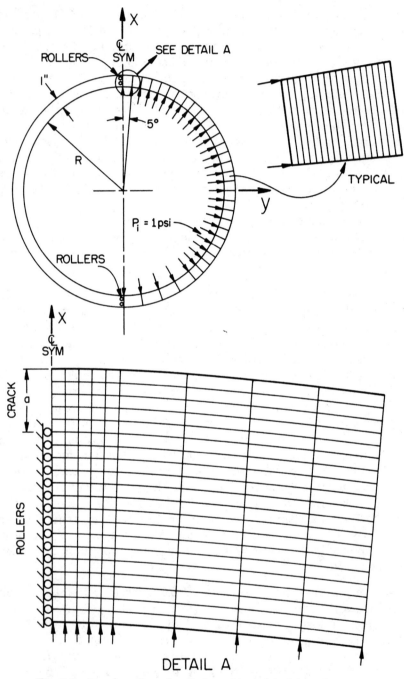

Fig. 2.7—A surface-flawed cylinder subjected to internal pressure.

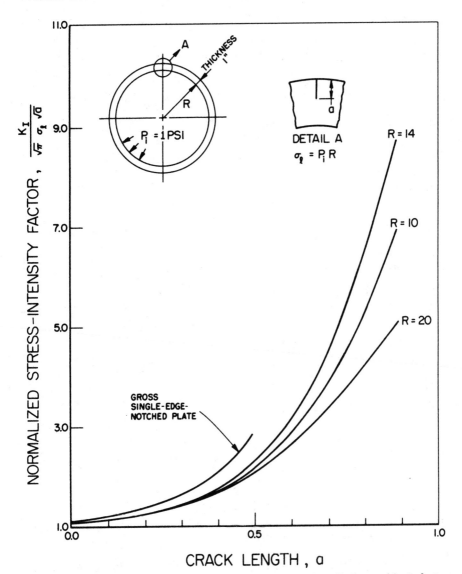

Fig. 2.8—Stress-intensity factor for a surface-flawed cylinder subjected to internal pressure.

mode deformation and u_y vanishes for the sliding-mode deformation. Thus, K can be determined from the u_y displacement on the crack surface, and K_{II} can be determined by the u_x displacement on the crack surface.

The above procedure was then used to determine stress-intensity factors in complex two-dimensional problems for studying crack-

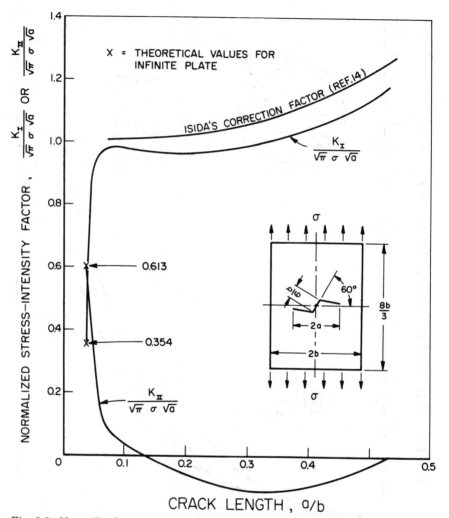

Fig. 2.9—Normalized stress-intensity factors for tension plates with initially slanted crack, $\alpha = 60$ deg.

propagation problems under bimodal crack deformation. Despite the abundance of literature on fatigue-crack-extension rates of the opening mode, no published information exists to date on the fatigue-crack-extension rate under the bimodal loading of opening and sliding modes. Studies by Erdogan and Sih considered the interaction of critical stress-intensity factors of the two opening and sliding modes, K_c and K_{IIc}, for a straight crack,[19] while Wu considered a combined-mode fracture criterion of a straight crack in anisotropic plates.[20] Quantitative investigations on curved fatigue cracks in brittle materials have not appeared in the literature. The lack of such studies can be attributed partly to the lack of analytical methods for determining K and K_{II} for a crack with a curved path. The stress-intensity factor for the curved crack noted in the previous paragraph was determined by the use of finite-element analysis and is shown in Fig. 2.9.

For $\alpha = 60$ deg, the two stress-intensity factors, K and K_{II}, of the crack prior to the growth of the elbow can be determined by eqs. (2.31) and (2.32), together with the stress equations of transformation as

$$K = \sqrt{\pi}\ \sigma \sqrt{a \sec \alpha}\ \cos^2 \alpha$$
$$K_{II} = \sqrt{\pi}\ \sigma \sqrt{a \sec \alpha}\ \sin \alpha \cos \alpha \qquad (2.37)$$

These theoretical values of K and K_{II} are also shown in Fig. 2.9 as a check on the accuracy of stress-intensity factors which are determined numerically.

The two modes of stress intensities, K and K_{II}, show a sharp change as the crack turns sharply from its initial orientation of α. Experimentally, these sharp turns were observed to occur immediately after the crack extends under cyclic loading. This abrupt discontinuity in crack path is approximated by the rapid but continuous changes in stress-intensity factor, as shown in Fig. 2.9. As the crack turns, the opening mode of stress-intensity factor, K, climbs rapidly to approach K of the horizontal crack. Likewise, K_{II} drops immediately to an order of magnitude of smaller value. Presumably, these somewhat gradual transitions in K and K_{II} are caused by averaging effects of the finite-element method when applied to the analysis of a discontinuous crack path such as considered here. These limited results suggest that a fatigue crack which is initially under bimodal loading propagates immediately in the direction of maximum K value, with the K_{II} value reducing to its minimum value.

Using the above stress-intensity factors K and K_{II}, the crack-propagation rate, $\Delta a/\Delta N$, was plotted against the maximum value of K following the fracture-mechanics approach to fatigue as initially proposed by Paris.[21] Figure 2.10 shows these results with the corresponding results for horizontal cracks (or $\alpha = 0$ deg). Although the results of crack-propagation rate ($\Delta a/\Delta N$) in slanted and unslanted cracks appear almost parallel in Fig. 2.10, it should be noted that the crack length referenced, Δa, is the projected length of the crack and, thus, the actual length of crack is initially much larger than its projected length. Toward the final stage of crack propagation, the actual and projected lengths of crack propagation become very close to each other. It can be seen from these results that, for a given applied stress-intensity-factor variation of K, say 11.5 ksi $\sqrt{\text{in}}$. in Fig. 2.10, the presence of K_{II}, no matter how small, will increase the crack-

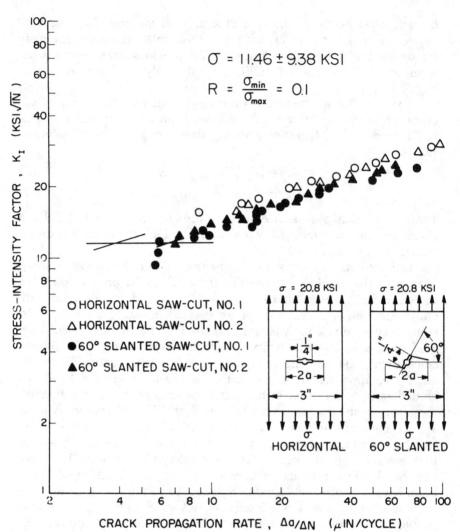

Fig. 2.10—Crack-propagation rate for 7075-T6 tension plate with initially slanted crack, $\alpha = 60$ deg.

propagation rate, $\Delta a/\Delta N$, significantly. Additional test results for $\alpha = 45$ deg and 90 deg are described in Ref. 22.

2.3 THREE-DIMENSIONAL PROBLEMS IN FRACTURE MECHANICS

Circular Crack

Since the famous papers on the penny-shaped crack by Sach[23] and Sneddon,[24] and by Green and Sneddon on the elliptical crack[25] in an infinite solid subjected to uniaxial tension load perpendicular

to the crack surface, a number of studies involving circular and elliptical crack with other loading conditions have been published. Unfortunately, these studies have limited application to practical problems where the effect of finite structural dimensions must be
accounted for. A typical example is the problem of surface flaw
shown in Fig. 2.11 which is the commonest type of crack and for
which there exist no analytical solutions to this date.

Recently, Thresher and Smith computed the stress-intensity factor of a part-through circular segment in a plate where the semielliptical flaw in Fig. 2.11 is replaced by a portion of a circular arc.[26]
This analysis further extends the procedure used in Ref. 27, where a
circular crack with sinusoidally varying pressure distribution which
was represented by a generalized double-Fourier series was considered. By applying an alternating technique between this solution and
Love's solution[28] for a half space in Ref. 27, Smith obtained numerically the stress-intensity factor along the crack periphery of a semi-

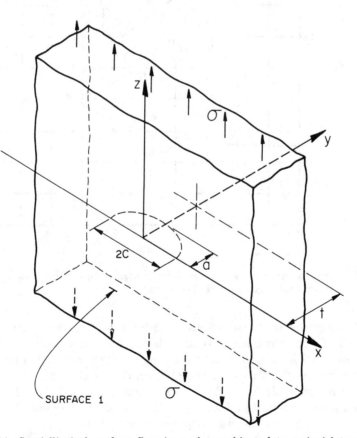

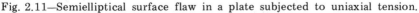

Fig. 2.11—Semielliptical surface flaw in a plate subjected to uniaxial tension.

circular surface flaw in a semi-infinite plate subjected to uniaxial tension as shown in Fig. 2.12. This figure shows that the stress-intensity factor increases as the crack approaches a free surface, but there is considerable doubt about the accuracy of this result as well as the singularity of $1/\sqrt{r}$ in the region where the crack periphery intersects the free surface.[29] The average value of the increase of stress-intensity factor along the entire crack periphery is about 8–9 percent, which agrees with the 10-percent increase estimated by Irwin.[30]

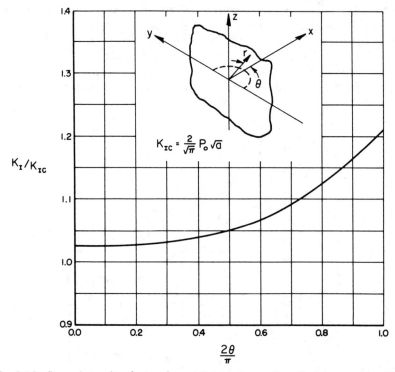

Fig. 2.12—Stress-intensity factor for a semicircular surface flaw in a semi-infinite solid subjected to uniaxial tension.

Among the other solutions described in Ref. 27 is the stress-intensity factor for a semicircular surface flaw in a beam in bending as shown in Fig. 2.13. It is interesting to note from this figure that the stress-intensity factor at the bottom of the flaw does not vanish when the crack tip is at the neutral axis or $a/c = 1.00$. Thus, the crack tip can penetrate farther into the compression region of the unflawed beam before the crack tip can no longer open up.

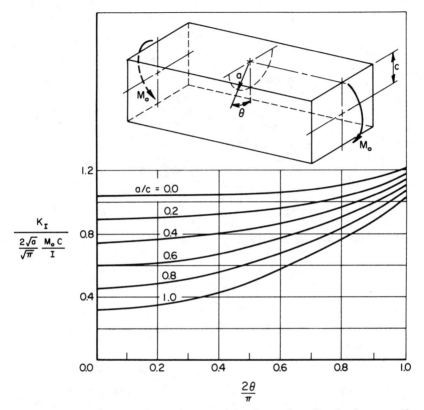

Fig. 2.13—Stress-intensity factor for a semicircular surface flaw in the top surface of a beam in pure bending.

Elliptical Crack

The results described above, as well as the new results in Refs. 31 and 32, are limited to flaw shapes of circles or segments of circles. Actual flaws, however, are closer to an ellipse in shape and, therefore, in the following, a brief description of the little available theoretical background on elliptical flaw will be given.

For an elliptical crack, as shown in Fig. 2.14, in an infinite solid and subjected to a pressure distribution of

$$p(x,y) = \sum_{i,j} A_{ij} x^i y^j \tag{2.38}$$

the stress function can be represented as

$$\phi = \sum_{i,j} c_{ij} \frac{\partial^{i+j} V^{(i+j+1)}}{\partial x_i \partial y_j} \tag{2.39}$$

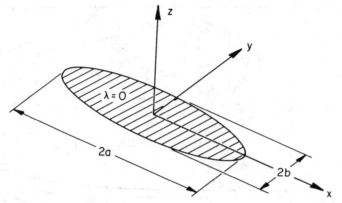

Fig. 2.14—Elliptical crack in an infinite solid.

where V is a potential function represented as

$$V^{(n)} = \int_{\lambda}^{\infty} \frac{\omega^n(s)}{\sqrt{Q(s)}} ds$$

$$\omega(s) = 1 - \frac{x^2}{a^2 + s} - \frac{y^2}{b^2 + s} - \frac{z^2}{s}$$

$$Q(s) = s(a^2 + s)(b^2 + s) \qquad (2.40)$$

λ, and also μ and ν, are coordinates of an ellipsoidal coordinate system and are roots of $\omega(s) = 0$. The crack surface is then represented as $\lambda = 0$.

This stress function satisfies the boundary conditions that

$$\sigma_{zz} = -\frac{\partial^2 \Phi}{\partial z^2}\bigg]_{\lambda=0} = \frac{p(x,y)}{2G} \quad \text{for} \quad \frac{x^2}{a^2} + \frac{y^2}{b^2} \leq 1 \quad \text{and} \quad z = 0$$

$$w = \frac{\partial \phi}{\partial z} = 0 \qquad\qquad \text{for} \quad \frac{x^2}{a^2} + \frac{y^2}{b^2} \geq 1 \quad \text{and} \quad z = 0$$

$$\sigma_{yz} = \sigma_{xz} = 0 \qquad\qquad \text{for} \quad z = 0 \qquad\qquad (2.41)$$

In addition, all stress and displacement components vanish at distances far away from the crack.

For a regular distribution of applied pressure on the crack surface, the stresses in the vicinity of the crack tip are shown to possess a $1/\sqrt{r}$ singularity with the dominant state of stress near the crack border being that of plane strain.[7,33] Thus, the stress-intensity factor derivable from the local state of stress is that of plane-strain mode I stress-intensity factor of

$$K_I = \frac{8G}{ab}\sqrt{\frac{\pi}{ab}}(a^2 \sin^2 \theta + b^2 \cos^2 \theta)^{1/4}\left\{ C_{00} + \frac{C_{10} \cos \theta}{a} \right.$$

$$+ \frac{C_{01} \sin \theta}{b} - \frac{4 C_{20} \cos^2 \theta}{a^2} + \frac{C_{11} \cos \theta \cdot \sin \theta}{ab}$$

$$- \frac{4 C_{02} \sin^2 \theta}{b^2} - \frac{4 C_{30} \cos^3 \theta}{a^3} - \frac{4 C_{21} \cos^2 \theta \cdot \sin \theta}{a^2 b}$$

$$- \frac{4 C_{12} \cos \theta \cdot \sin^2 \theta}{ab^2} - \frac{4 C_{03} \cdot \sin^3 \theta}{b^3} + \cdots \qquad (2.42)$$

where θ is the angle in the parametric representation of ellipse of

$$x = a \cos \theta \qquad y = b \sin \theta \qquad z = 0$$

In theory, the above series can be carried out to many terms but the derivations are lengthy and, thus, only 10 terms are derived in Ref. 33. These terms correspond to the coefficients of $C_{00}, C_{01}, \ldots$ C_{03} and are related through eq. (2.41) to the pressure distribution of

$$p(x,y) = A_{00} + A_{10}x + A_{01}y + A_{20}x^2 + A_{11}xy + A_{02}y^2$$
$$+ A_{30}x^3 + A_{21}x^2y + A_{12}xy^2 + A_{03}y^3 \qquad (2.43)$$

The known solutions can now be derived as special cases of the stress-intensity factor represented by eq. (2.42). In the following, some of these known solutions will be derived.

Elliptical crack in an infinite solid subjected to uniaxial tension. The tension problem of a crack in an infinite solid can be replaced by a pressurized crack in an infinite solid.[9] Thus, consider an elliptical crack subjected to a constant applied pressure of $\sigma_{zz} = -p_o$. From eq. (2.43), $A_{00} = -p_o$ and $A_{10} = A_{01} = \cdots = A_{03} = 0$. As a result,

$$C_{00} = \frac{p_o ab^2}{8GE(k)} \quad \text{and} \quad C_{01} = C_{10} = \cdots = C_{12} = C_{03} = 0 \quad (2.44a)$$

where $E(k)$ is the complete elliptical integral of the second kind and

$$k^2 = 1 - \frac{a^2}{b^2} \qquad (2.44b)$$

The stress-intensity factor can then be derived from eq. (2.42) as

$$K_I = \frac{\sqrt{\pi}}{E(k)}\sqrt{\frac{b}{a}}\, p_o\, (a^2 \sin^2 \theta + b^2 \cos^2 \theta)^{1/4} \qquad (2.45)$$

which was also derived by Irwin[30] based on the solution by Green and Sneddon.[25]

Elliptical crack in an infinite solid subjected to pure bending. This problem can be replaced by a problem of a linearly varying pressure of $\sigma_{zz} = p_o x$ on the elliptical crack surface or $A_{10} = -p_o$ and $A_{00} = A_{01} = A_{20} = \cdots = A_{03} = 0$. Then

$$C_{10} = \frac{p_o a^3 b^2 k^2}{8G[(1 - 2k^2)E(k) - k'^2 K(k)]}$$

$$C_{00} = C_{01} = C_{20} = \cdots = C_{12} = C_{03} = 0 \qquad (2.46)$$

where $K(k)$ is the complete elliptical integral of the first kind

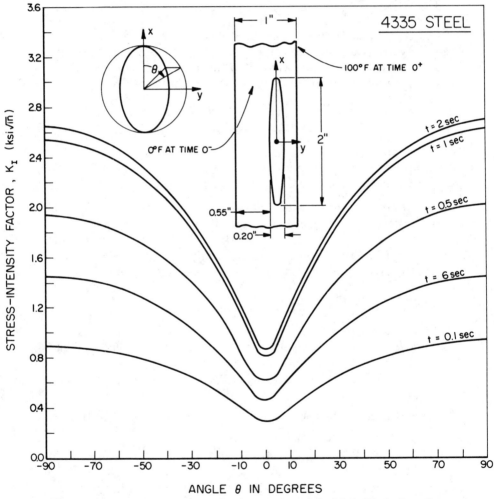

Fig. 2.15—Transient stress-intensity factor of an embedded elliptical crack.

$$k' = \frac{a}{b} \qquad (2.47)$$

The stress-intensity factor becomes

$$K_I = \frac{p_o a k^2 \sqrt{\pi}}{[(1 - 2k^2)E(k) - k'K(k)]}$$

$$\cdot \sqrt{\frac{b}{a}} \cos \theta \, (a^2 \sin^2 \theta + b^2 \cos^2 \theta)^{1/4} \qquad (2.48)$$

which was also derived by Kassir and Sih.[34]

Elliptical crack in a thick plate subjected to sudden heating on one surface. The procedure has been used in determining the transient stress-intensity factor of an elliptical crack embedded in a thick plate resting at $0°F$ and which was then subjected to sudden heating to $100°F$ of one surface. The transient thermal stress in the unflawed plate was computed and then the tension stresses at the flaw location were freed by pressurizing the elliptical crack with a variable pressure equal to the thermal stresses. The assumption of thick plate precludes the interaction in stress distribution between the two free surfaces of the plate and the pressurized crack. Figure 2.15 shows the transient stress-intensity factor along the crack periphery. While the maximum value of K_I is only 2 ksi $\sqrt{\text{in}}$. in this calculation, this value superimposed on the stress-intensity factor due to mechanical loading, such as pressure loading in a pressure vessel, could well elevate K_I beyond the fracture toughness of material at that temperature, and could cause catastrophic failure of the structure.

Semielliptical surface flaw. The problem of a semielliptical surface flaw as shown in Fig. 2.11 remains unsolved except for the recent work by Thresher[26] which is limited to high aspect ratio of $a/2c = 0.3$ and relatively shallow flaws with $a/t = 0.8$. Flaw shapes of $a/2c = 0.12$ to 0.2 with depth of $a/t = 0.9$ are probably the most critical problems in applied fracture mechanics without an adequate solution. Thus, a semiempirical formula for estimating this deep-flaw effect was derived in Ref. 35 through the judicious use of known solutions and was verified by a limited number of experimental results.

In addition, results from recently completed analytical work,[36] coupled together with portions of the empirical results in Ref. 35, were used to establish elastic magnification factors of higher accuracy for deep surface flaws, i.e., $a/t > 0.8$. These elastic magnification factors are shown in Fig. 2.16.

From this brief exposition on three-dimensional analytical technique, it can be seen that the state of the art is very limited at this

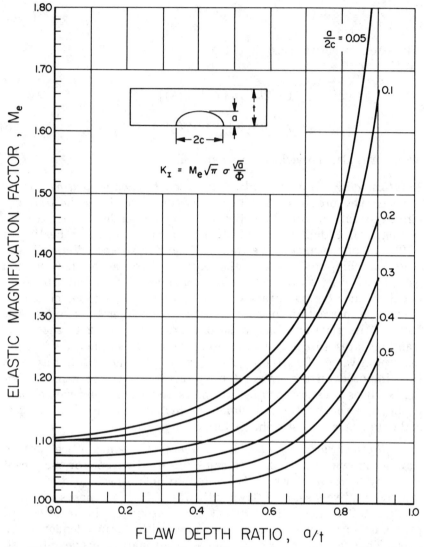

Fig. 2.16—Elastic stress-intensity magnification factor for a surface-flawed tension plate.

time. Unlike two-dimensional solutions in fracture mechanics, there exist very few analytical solutions for three-dimensional problems.

Finite-element Method

The finite-element method used in two-dimensional analysis is certainly usable in three-dimensional analysis within the limitation of

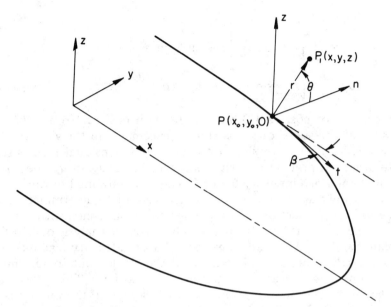

Fig. 2.17—Local-coordinate system for a curved crack.

present-generation computers. Since the complexities of three-dimensional finite-element method are multiplied manyfold over two-dimensional finite-element method, few applications of the procedure to three-dimensional problems in fracture mechanics are known at this time.

For the determination of stress-intensity factor which is in the state of plane strain, except for possibly in regions close to the free surface,[29] the opening-mode stress-intensity factor can be represented in terms of a local coordinate system t-n-z, which is tangential and normal to the crack front as shown in Fig. 2.17, as[33]

$$\sigma_{tt} = 2\nu \frac{K_I}{\sqrt{2\pi r}} \cos \frac{\theta}{2} + \mathcal{O}(\sqrt{r})$$

$$\sigma_{nn} = \frac{K_I}{\sqrt{2\pi r}} \cos \frac{\theta}{2} \left[1 - \sin \frac{\theta}{2} \sin \frac{3\theta}{2}\right] + \mathcal{O}(\sqrt{r})$$

$$\sigma_{zz} = \frac{K_I}{\sqrt{2\pi r}} \cos \frac{\theta}{2} \left[1 + \sin \frac{\theta}{2} \cdot \sin \frac{3\theta}{2}\right] + \mathcal{O}(\sqrt{r})$$

$$\sigma_{tn} = \mathcal{O}(\sqrt{r})$$

$$\sigma_{tz} = \mathcal{O}(\sqrt{r})$$

$$\sigma_{nz} = \frac{K_I}{\sqrt{2\pi r}} \sin \frac{\theta}{2} \cos \frac{\theta}{2} \cdot \cos \frac{3\theta}{2} + \mathcal{O}(\sqrt{r})$$

$$(2.49)$$

$$u_t = \theta \ (r)$$

$$u_n = \frac{K_I}{G} \sqrt{\frac{r}{2\pi}} \cos \frac{\theta}{2} \left[1 - 2\nu + \sin^2 \frac{\theta}{2} \right] + \mathcal{O}(r)$$

$$u_z = \frac{K_I}{G} \sqrt{\frac{r}{2\pi}} \sin \frac{\theta}{2} \left[2 - 2\nu - \cos^2 \frac{\theta}{2} \right] + \mathcal{O}(r) \qquad (2.50)$$

Comparison of eqs. (2.49) and (2.50) with eqs. (2.10) and (2.11) shows that the states of stress and displacements in the vicinity of a curved crack front are similar to the two-dimensional plane-strain state discussed previously. Thus, knowing either σ_{zz} or u_z, the opening mode of stress-intensity factor K_I can be determined readily.

The total-elastic-strain-energy procedure for K_I determination, as discussed in the section of two-dimensional finite-element analysis, cannot be used effectively in three-dimensional analysis due to the variability of K_I along the crack front. Such difficulty may be overcome by advancing a small segment of the crack front to determine the strain-energy-release rate at each local region. This procedure must then be repeated over the entire crack front if the variations in stress-intensity factor are to be determined. The greatly increased amount of computation time, however, makes this procedure impractical to use in three-dimensional analysis.

Recently, Marcal et al. developed a three-dimensional polar element which uses the crack tip as the focus.[37] This element has a rational strain singularity of $1/r$ but appears to represent well the $1/\sqrt{r}$ stress singularity in elastic analysis. Some preliminary results on the stress-intensity factors of three-dimensional through-the-thickness crack and semielliptical-surface-flaw problems are reported.

Another set of limited preliminary results on the stress-intensity factor for a surface flaw in a tension specimen was reported by Miyamoto and Miyoshi.[38] The finite element used in this analysis is the parallel pipe element obtained by merging six tetrahedral elements together. The crack-opening displacement necessary for stress-intensity-factor determination was obtained through a two-stage computation involving coarse-grid and fine-grid analysis of a quarter segment of the specimen, similar to the procedure described in the section of two-dimensional finite-element analysis. The elastic magnification factor for the deep surface flaw appears to be somewhat underestimated in this analysis.

2.4 LIMITATION OF ANALYTICAL TECHNIQUES AND THE ROLE OF EXPERIMENTAL MECHANICS

The analytical procedures described above show that many of the two-dimensional problems in linear fracture mechanics can be solved by the elegant analytical techniques of Muskhelishvili or the numerical method of finite-element analysis. There are, however, many practical problems in fracture mechanics involving plasticity, viscoelasticity, and dynamic elasticity where available analytical and numerical techniques are not advanced enough to replace traditional and newer experimental techniques in fracture mechanics.

As for three-dimensional problems in fracture mechanics, most problems in linear-elastic fracture mechanics remain unsolved mainly due to limitation in available analytical techniques and the computer capacity. Although experimental difficulties are also compounded in solving problems in three-dimensional fracture mechanics, experimental method appears to be the only general and more readily accessible procedure for analyzing three-dimensional problems in fracture mechanics at the present time.

In subsequent chapters, these experimental techniques in fracture mechanics will be discussed in detail.

REFERENCES

1. Irwin, G. R., and Wells, A. A., "A Continuum Mechanics View of Crack Propagation," *Metallurgical Review*, **10** (58), 223-270 (1965).
2. Griffith, A. A., "The Phenomena of Rupture and Flow in Solids," *Phil. Trans. of Royal Soc. of London*, **221**, 163-198 (1921).
3. Irwin, G. R., and Kies, J., "Fracturing and Fracture Dynamics," *Welding Jnl. Research Supplement* (Feb. 1952).
4. Orowan, E., "Fundamentals of Brittle Behavior of Metals," *Fatigue and Fracture of Metals*, John Wiley & Sons, 139-167 (1952).
5. Bueckner, H. F., "The Propagation of Cracks and Energy of Elastic Deformation," *Jnl. of Appl. Mech., Trans. of ASME*, 1225-1229 (Aug. 1958).
6. Irwin, G. R., "Fracture," *Handbuch der Physik*, **6**, 551-590 (1958).
7. Sih, G. C., and Liebowitz, H., "Mathematical Theories of Brittle Fracture," *Fracture* (editor, H. Liebowitz), 2, 67-190 (1968).
8. Rice, J. R., "Mathematical Analysis in the Mechanics of Fracture," *Fracture* (editor, H. Liebowitz), 2, 191-311 (1968).
9. Paris, P. C., and Sih, G. C., "Stress Analysis of Cracks," *Fracture Toughness Testing and Its Applications*, ASTM STP 381, 39-76 (1965).
10. Muskhelishvili, N. I., *Some Basic Problems of the Mathematical Theory of Elasticity*, translated by J. R. M. Radok, Noordhoff (1953).
11. Zienkiewicz, O. C., and Cheung, Y. K., *The Finite Element Method in Structural and Continuum Mechanics*, McGraw-Hill (1965).
12. Watwood, V. B., "The Finite Element Method for Prediction of Crack Behavior," *Nuclear Engrg. and Design*, 11, 323-332 (1969).
13. Deverall, L. L., and Lindsey, G. H., "A Comparison of Numerical Methods for Determining Stress Intensity Factors," ICRPG Paper (1969).
14. Kobayashi, A. S., Chiu, S. T., and Beeuwkes, R., "Elastic-Plastic State in a Plate with an Extending Crack," *Proc. of the Army Symposium on Solid Mechanics—Lightweight Structures* (1970).

15. Isida, M., and Itagaki, Y., "Stress Concentration at the Tip of a Central Transverse Crack in a Stiffened Plate Subjected to Tension," *Proc. of 4th U.S. Natl. Cong. of Appl. Mech.*, 955-970 (1962).
16. Tiffany, C. F., and Masters, J. N., *Applied Fracture Mechanics*, ASTM STP 381, 249-278 (1965).
17. Folias, E. S., "On the Theory of Fracture of Curved Sheet," *Jnl. of Engrg. Fract. Mech.* 2, 151-164 (1970).
18. Gross, B., Srawley, J. E., and Brown, W. F., *Stress Intensity Factor for a Single-Edge-Notch Tension Specimen by Boundary Collocation*, NASA TN D-2395 (Aug. 1964).
19. Erdogan, F., and Sih, G. C., "On the Crack Extension in Plates under Plane Loading and Transverse Shear," *Jnl. of Basic Engrg.*, Trans. of ASME, 519-527 (Dec. 1963).
20. Wu, E. M., "Application of Fracture Mechanics to Anisotropic Plates," *Jnl. of Appl. Mech.*, Trans. of ASME, 34, Series E (4) 967-974 (Dec. 1967).
21. Paris, P. C., "The Fracture Mechanics Approach to Fatigue," *Proc. of 10th Sagamore Conf.*, Syracuse University Press, 107-132 (1965).
22. Iida, S., and Kobayashi, A. S., "Crack Propagation Rate in 7075-T6 Plates under Cyclic Tensile and Transverse Shear Loadings," *Jnl. of Basic Engrg.*, Trans. of ASME, 91, Series D (4), 764-769 (Dec. 1964).
23. Sach, R. A., "Extension of Griffith Theory of Rupture to Three Dimensions," *Proc. of Phys. Soc.*, London, 58, 729-736 (1946).
24. Sneddon, I. N., "The Distribution of Stress in the Neighborhood of a Crack in an Elastic Solid," *Proc. of the Royal Soc.*, London, Series A, 187 (1946).
25. Green, A. E., and Sneddon, I. N., "The Stress Distribution in the Neighborhood of a Flat Elliptical Crack in an Elastic Solid," *Proc. of Cambridge Philosophical Soc.*, 46 (1956).
26. Thresher, R. W., and Smith, F. W., "Stress Intensity Factors for a Surface Crack in a Finite Solid," *Jnl. of Appl. Mech.*, Trans. of ASME, 95, 195-200 (Mar. 1972).
27. Smith, F. W., Emery, A. F., and Kobayashi, A. S., "Stress Intensity Factors for Semi-Circular Cracks, Part II—Semi-Infinite Solid," *Jnl. of Appl. Mech.*, Trans. of ASME, 34, Series E, 953-959 (Dec. 1967).
28. Love, A. E. H., "On the Stress Produced in a Semi-Infinite Solid by Pressure on Part of the Boundary," *Phil. Trans. of the Royal Soc.*, London, 228, Series A, 378-395 (1929).
29. Sih, G. C., "Three-Dimensional Stress-State in a Cracked Plate," *Proc. of the Air Force Conf. on Fatigue and Fracture of Aircraft Structures and Materials* (Sept. 1970).
30. Irwin, G. R., "The Crack Extension Force for a Part-Through Crack in a Plate," *Jnl. of Appl. Mech.*, Trans. of ASME, 29, 651-654 (Dec. 1962).
31. Smith, F. W., and Alavi, M. J., "Stress-Intensity Factors for a Penny-Shaped Crack in a Half Space," *Intl. Jnl. of Engrg. Fract. Mech.*, 3, 241-254 (Oct. 1971).
32. Smith, F. W., and Alavi, M. J., "Stress-Intensity Factors for a Part-Circular Surface Flaw," *Proc. of 1st Intl. Conf. on Pressure Vessels and Pipe Lines*, Delft, Holland (Aug. 1969).

33. Shah, R. C., and Kobayashi, A. S., "Stress-Intensity Factor for an Elliptical Crack under Arbitrary Normal Loading," *Intl. Jnl. of Engrg. Fract. Mech.*, 3, 71–96 (July 1971).
34. Kassir, M. K., and Sih, G. C., "Three-Dimensional Stress Distribution around an Elliptical Crack under Arbitrary Loadings," *Jnl. of Appl. Mech.*, Trans. of ASME, 33, 601–611 (Sept. 1966).
35. Kobayashi, A. S., and Moss, W. L., "Stress-Intensity Magnification Factors for a Surface-Flawed Tension Plate and Notched Round Bar," *Proc. of 2nd Intl. Conf. on Fract.*, Brighton (1969).
36. Shah, R. C., and Kobayashi, A. S., "Stress-Intensity Factors for an Elliptical Crack Approaching the Surface of a Semi-Infinite Solid," to be published in the *Intl. Jnl. of Engrg. Fract. Mech.*
37. Levy, N., Marcal, P. V., and Rice, J. R., "On the Development of Finite-Element Computational Methods for Three Dimensional Crack Analysis," Heavy Section Steel Technology 4th Annual Information Meeting, Paper No. 14, Oak Ridge National Laboratory (Apr. 1970).
38. Miyamoto, H., and Miyoshi, T., "Analysis of Stress-Intensity Factor for Surface-Flawed Tension Plate," *Proc. of the Symposium on High Speed Computing of Elastic Structures (IUTAM)*, Liege, Belgium (Aug. 1970).

3

ACOUSTIC-EMISSION TECHNIQUES

by

H. L. Dunegan and D. O. Harris
Dunegan/Endevco
Livermore, Calif.

3.1 SYMBOLS

a	= radius of penny-shaped crack, crack length in plane specimen
a_0	= initial crack length
ΔA	= incremental area swept out by crack
B	= plate thickness
C	= proportionality constant in crack-growth law [see eq. (3.10)]
D	= proportionality term [see eq. (3.3)]
D'	= proportionality constant [see eq. (3.1)]
D''	= proportionality term [see eq. (3.4)]
E	= Young's modulus
F	= tensile load on fracture toughness specimen
G	= see eq. (3.9)
Σg	= summation of amplitude of emission pulses
K	= mode I stress-intensity factor
$\hat{K}$	= range of stress-intensity factor ($= K_{\max} - K_{\min}$)
K_c	= critical stress-intensity factor for onset of rapid crack extension
K_{ISCC}	= threshold stress-intensity factor for stress-corrosion cracking
n	= number of fatigue cycles
N	= number of acoustic-emission counts
N_t	= total number of acoustic-emission counts observed during proof loading
q	= exponent in crack-growth law [see eq. (3.10)]
r_y	= plastic-zone size
t	= time
V_p	= volume of plastically deformed material
W	= width of wedge-opening-loading specimen

α $= a/W$
α_0 $= a_0/W$
$\eta(\alpha)$ = value of integral [see eq. (3.13)]
σ_p = proof stress
σ_w = working stress during fatigue loading
σ_{ys} = engineering yield stress

3.2 INTRODUCTION

Acoustic emission is the term applied to the low-level stress waves emitted by a material when it is deformed, either by an external stress or internal process, such as a martensitic-phase transformation.

One of the first known uses of acoustic emission dates back to the middle of the last century. At that time, miners began digging tunnels in the Sierras in search of gold. Most of the miners were inexperienced in mining engineering and were, therefore, not aware of many of the safety precautions prevailing in the mining industry at that time. As a result of their ignorance, they soon divided themselves into two groups. Those who learned to heed the creaking of the mine timbers and left the mine became known as the "quick," while those who passed off this acoustic emission as a mere scientific curiosity rarely earned the title of "Old Timer."

This lesson of 100 years ago is just as applicable to modern technology as it was then when considering that today's space-age high-strength materials are much less tolerant to the presence of flaws than were the noisy structures of that era. While not as tolerant as wood, neither are they as noisy, for the frequency and level of sound waves emitted by most metals under stress are not in the range of detection of the human ear. There are exceptions, such as the emission from tin or zinc when grossly deformed, and the pop-in emission from fracture specimens when the crack front becomes momentarily unstable, and then quickly arrests. Thus, acoustic-emission signals from modern structures are normally detected and recorded with state-of-the-art transducers and instrumentation. The purpose of this chapter is to describe the mechanisms responsible for acoustic emission, the instrumentation and techniques available for recording the signals, and the application of this information to the field of fracture mechanics.

Historical Review

Acoustic emission from rocks. The impetus for engineering research is often promoted by a catastrophe and the loss of human lives. This was one of the factors that instigated an investigation of the noise in deep mines to determine if a useful procedure could be developed to predict the collapse of tunnel walls. As early as 1923, Hodgson[1] proposed the use of subaudible noise in the prediction of rockbursts and earthquakes, but it was not until the 1940s that acoustic techniques were applied to indicate the approach of rockbursts in mines.[2,3] In the early 1950s, Russian investigators used listening apparatus to forecast outbursts of coal and gas in mines,[4] and these techniques are in wide use in Russia today. Reference 5 provides a good review of the use of acoustic emission in rock-mechanics studies. More recent work by Scholz is included in Refs. 6–8.

Acoustic emission from metals. Allusions to the occurrence of sound emitted by metals during plastic deformation is traceable in the literature at least back to the year 1928. The early observations were largely restricted to twinning of large single crystals, *tin cry*, and twinning or crystal transformations on a comparatively large scale within the specimens studied. B. H. Schofield gives a historical bibliography in one of his research reports.[9]

It was not until 1950 that a serious study of acoustic emission from materials other than rocks was started. Joseph Kaiser[10] performed the first experiments on conventional engineering materials and demonstrated that acoustic emission accompanied permanent deformation processes in metals. Following Kaiser's work, Schofield[9] performed an extensive investigation into the surface and volume effects of both single-crystal and polycrystalline metals. His most important conclusions were that unpinning of groups of dislocations was the primary mechanism responsible for acoustic emission in nontwinning metals, and that both polycrystalline and single-crystal materials emit acoustic-emission signals when deformed. This study refuted the interpretation of Kaiser that the source of acoustic emissions was frictional rubbing of grains against each other. Tatro initiated investigations into the acoustic-emission phenomenon in 1956. He and his students worked primarily in the nominal elastic range of aluminum and at very low strain rates. He and Liptai[11] also performed experiments on anodized aluminum to determine the effects of surface treatment on the acoustic-emission behavior. J. R. Frederick at the University of Michigan has carried on investigations in acoustic emission for several years. One of his primary fields of interest has been in the area of fatigue. Two Ph.D. theses[12,13] attest to the results of this work.

The earliest work concerned with the emission from a flawed specimen appears to be that conducted by Romine.[14] Jones and

Brown[15] and Green et al.[16] were also early workers in this field. However, all of these investigators used relatively insensitive transducers. Schofield[17] used sensitive transducers in an investigation of the emission from a structure containing a flaw in the form of a machined notch. The authors[18] were the first to use sensitive transducers in the study of emission from structures containing realistic flaws such as fatigue cracks.

Most of the work prior to 1964 was conducted on strictly a laboratory basis with carefully controlled conditions. Since most of the early investigators chose to work in the frequency range used by Kaiser (below 60 Khz), elaborate soundproof chambers and loading devices were necessary to prevent background noise from interfering with the experiments. The authors[19] were the first (to our knowledge) to extend experiments into the hundred-kilocycle and megacycle range to eliminate extraneous noise. The early experiments demonstrated that successful data could be obtained at the higher frequencies, thereby opening the door for practical application of acoustic-emission testing.

There are presently many laboratories, both domestic and foreign, that are active in acoustic-emission research. Some are concerned primarily with materials research, others with developing nondestructive test methods, and still others in applying acoustic emission to fracture-mechanics studies. It is the latter topic that will receive the major emphasis in this report, although it should become obvious that applications in fracture mechanics lead directly to applications to the nondestructive testing of structures.

3.3 INSTRUMENTATION

General

The individual events responsible for acoustic emission from metals are very short lived. It appears that these events have rise times in the range 10^{-4} to 10^{-8} sec; so, an acoustic-emission pulse contains a very wide range of frequency components, extending up into the megacycle range. Hence, any transducer sensitive enough to the pressure levels present will be responsive, regardless of its frequency response. Thus, one can choose from a wide band of frequencies the one most suitable for the test in mind. The distance between the acoustic-emission source and the receiving transducer, and the background noise level are primary factors in selecting a particular frequency band. Too high a frequency is not desirable, because the high-frequency components of the pulse are more severely attenuated than lower frequencies as they travel through the material. Too low a frequency is also not desirable, since background laboratory noise is then a problem. The optimum-frequency band is usually a compromise between these two opposing factors.

Low-frequency transducers (geophones) were used by Cook[20] in the frequency range 10 to 200 Hz for successfully detecting and locating rock bursts in deep mines. The detection range at the low frequency is 100 ft and more. The heterogeneous nature of rock and soil would prohibit recording the very high-frequency components present in the pulses, because of the high attenuations—unless the transducer was located very near the source. Hutton[21] performed acoustic-emission experiments on an operating nuclear reactor to determine if acoustic-emission signals from a growing crack could be detected above the background noise of the reactor pumping system. His experiments indicated that a detection system operating in the megacycle range would be necessary to eliminate unwanted noise. These two examples show the broad-band nature of acoustic-emission pulses and the complications and limitations that are present with a given system. The authors have found that the frequency range of 100 to 300 Khz is sufficiently high to eliminate most background noise associated with fracture-mechanics studies, while being low enough that attenuation effects are unimportant. A further advantage of this frequency is that the data can be easily recorded on magnetic tape for later analysis.

Detailed examination of the individual events responsible for acoustic-emission is complicated by several factors. Dispersion of the energy from an event will occur, depending on the boundaries of the specimen. The boundaries will also result in mode conversion, so that shear, longitudinal, and Rayleigh waves will be present due to a single event. Since many events occur in the process of stressing a specimen or structure, a very complex situation exists, which makes it difficult to obtain frequency information in the pulses that may be related to the internal processes involved, such as slip, twinning, microcrack formation, etc. Further complications arise due to mechanical filtering of the pulses by the structure, and the limitations imposed by the bandwidth–sensitivity product of the detecting transducer.

Transducers

Most investigators applying acoustic-emission techniques to metals have been and are still using a piezoelectric crystal placed in direct contact with the specimen or structure. Early investigators[9,10] used piezoelectric materials such as ammonium dihydrogen phosphate (ADP), Rochelle salt, and quartz as their acoustic-emission-transducer material. Development of the ferroelectric ceramics in recent years has led to superior detection ability. PZT-5 (Clevite Corporation) is one such material that has proved very useful for acoustic-emission transducers.

Jones and Brown[15] used a phonograph pickup to detect the plane strain pop-in of a cracked specimen. Green[16,22] has successfully used accelerometers for detecting acoustic emission from fiberglass composites and prestressed concrete. The authors have used the PZT material in the form of disks from $\frac{1}{2}$ in. to 1 in. in diameter, polarized in the thickness-expander mode. These are chosen so that the fundamental resonance falls in the range 100 to 300 Khz, and the associated electronics is bandpass filtered around this resonance. It has been observed that a frequency analysis of the pulses observed simply mirrors the response of the transducer and tells little of the true form of the initial pressure pulse, with the exception of its length. Thus, it is evident that an incoming acoustic-emission pulse sets the transducer to "ringing" at one of its resonant frequencies.

Electronics

The numerous different acoustic-emission systems have been used by investigators in the field; all have one feature in common, a low-noise preamplifier. A wide variety of components has been used for the remainder of the system. Some investigators record the pulses directly on a strip-chart recorder, others use oscilloscope readout, while most resort to electronic counters for measuring the number of counts as a function of time, or other test parameter of interest. Engle[23] and Fisher and Lally[24] used a rectification technique that gave one count for each acoustic-emission pulse. Dunegan and Tatro[25] used the unrectified pulse with a counter so that each oscillation of the transducer at its resonant frequency that results in a peak voltage above a preset trigger level of the counter is counted. In this manner, a large-amplitude event will give more counts than a small one because the larger the event, the larger the number of oscillations of the transducer before they decay below the trigger level. Thus, it is seen that some measure of the amplitude of the events is obtained by analyzing the emission in this manner. Figure 3.1 illustrates the difference between a rectified and unrectified acoustic-emission pulse. The voltage trigger level is indicated on the trace of the unrectified pulse, and this pulse would give approximately 15 counts. The rectified pulse would give one.

Figure 3.2 is a block diagram of a typical electronics system. After amplification, the acoustic emission signals are bandpass fil-

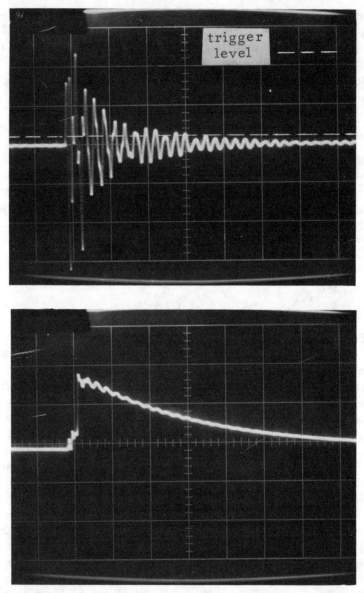

Fig. 3.1—Oscilloscope traces of a typical acoustic-emission pulse showing the difference between a rectified and unrectified pulse.

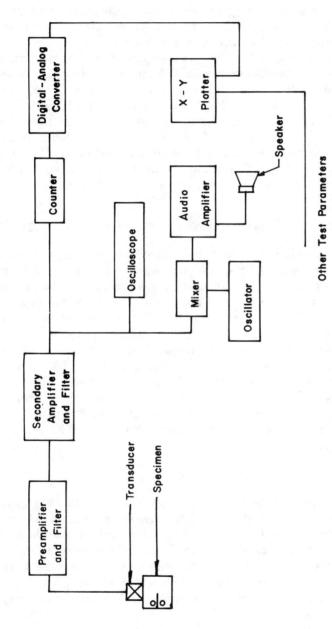

Fig. 3.2—Block diagram of instrumentation setup for acoustic-emission measurements.

tered and inserted into a counter, which can be set to either sum all counts or to provide a count rate. A digital to analog converter is used to obtain a DC voltage proportional to the number of counts for display on an X-Y plotter. The acoustic emission data can then be plotted directly as a function of other test parameters.

The acoustic emission signals are also mixed with a local oscillator (set near the transducer resonant frequency), low pass filtered and amplified by an audio amplifier. The audio amplifier output is connected to a speaker, so that an audio measure of the acoustic emission activity is obtained during the test. Several other accessories can also be included, such as an oscilloscope, digital printer, and analog tape recorder. Overall system gains of 60 to 100 db are used, depending on the amplitude of the acoustic emission events expected during the test.

3.4 USE OF ACOUSTIC EMISSION FOR FLAW DETECTION

As mentioned in earlier sections, acoustic emission from metals is related to permanent-deformation processes, such as plastic deformation. The onset of plastic deformation is controlled by the presence of stress raisers, such as notches and cracks. If no stress raisers are present, plastic flow will not begin until the nominal stresses are close to the general yield level. However, if stress raisers are present, plastic deformation will begin at nominal stress levels considerably below general yield, which will result in acoustic emission at low stresses. Hence, the presence of stress raisers greatly affects the nominal stress level at which acoustic emission begins, and monitoring for acoustic emission provides a technique for the detection in structures of the presence of stress raisers, such as cracks and other types of flaws.

Emission from Unflawed Materials

Some background on the acoustic emission from unflawed specimens is necessary when discussing the use of acoustic emission for flaw detection. Since the primary interest of this chapter is in materials containing cracks, only a brief review of emission from unflawed materials will be presented. Reference 25 provides a more complete discussion of emission from flaw-free materials.

Acoustic emission from unflawed materials generally consists of two components:

1. Low-level continuous: This component is generally continu-

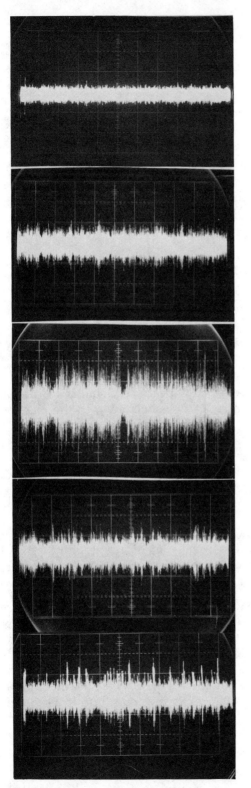

background

preyield

yield

postyield

near failure

Fig. 3.3—Oscilloscope traces of acoustic emission from a 7075-T6 aluminum
tensile specimen at different stages of strain.

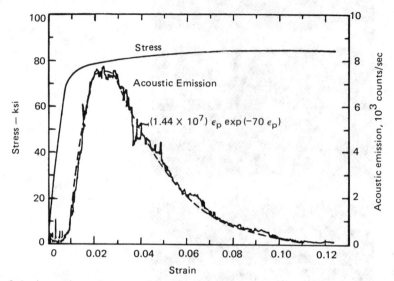

Fig. 3.4—Acoustic emission and stress vs. strain for a 7075-T6 aluminum tensile
specimen. Dashed curve is a fit of Gilman's[26] theoretical expression
for mobile dislocation density vs. plastic strain. (From Ref. 29)

ous. On an oscilloscope, it appears very similar to the background electrical noise. It is associated with the plastic deformation occurring at relatively small plastic strain. As the
load is increased in a tensile test, this component increases in
amplitude—this increase looks like a growth of noise. This
component decreases with increasing strain after yielding
occurs and is replaced by the second component at strains
near failure.

2. High-level burst: This component is not continuous, but
   occurs in bursts. It is usually of higher amplitude than the
   component discussed above, and is associated with microcrack
   formation that occurs at large plastic strains.

Figure 3.3 shows the oscilloscope traces of these two components
of acoustic emission as observed from unflawed material. This figure
shows the maximum in continuous emission observed near first yielding, and the increase in burst-type emission at larger plastic strains.

Figure 3.4 shows the acoustic-emission count rate as a function
of strain for a tensile test conducted on 7076-T6 aluminum. This
figure shows the characteristic peak in the count rate at plastic strain
near the yield strain. Also shown in this figure is a fit of Gilman's[26]
equation for the mobile dislocation density as a function of plastic
strain. The excellent fit of Gilman's equation to the acoustic

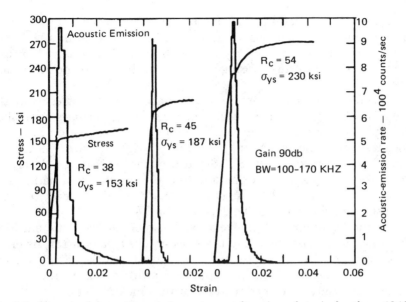

Fig. 3.5—Stress and acoustic-emission rate as a function of strain for three 4340 tensile specimens heat treated to different yield strengths.

emission data is further evidence of the close association between acoustic emission and the dislocation movement associated with plastic deformation in metals. Acoustic emission from 4340 steel tensile specimens heat treated to various strength levels is presented in Fig. 3.5. These results again show the characteristic shape of the count rate–strain curve, with the peak near the yield strain. This characteristic shape has been observed in many other metals by the authors, independently, by Fisher and Lally[24] at United States Steel and also by many other investigators.

Another important feature of acoustic emission in general is its irreversibility. If a material is loaded to a given stress level, and then unloaded, no emission will be observed upon reloading until this previous load has been exceeded. This is an expected result, since acoustic emission is closely associated with plastic deformation, which is also irreversible. This irreversibility of acoustic emission has important practical implications since it can be used in the detection of the subcritical growth of flaws, such as fatigue-crack growth, as will be discussed later. It can also be used in the development of passive peak pressure and acceleration transducers.[27]

Emission from Flawed Materials

The earliest studies of acoustic emission from flawed materials were concerned with the detection of pop-in.[14,15] This requires relatively insensitive transducers, because pop-in is usually quite noisy. This application of acoustic emission will be covered in a later section, and this section will deal with the emission observed from a flawed material during a rising-load test, with little or no subcritical flaw growth.

As discussed elsewhere in this monograph, the stresses in the vicinity of a crack tip in an elastic material are completely controlled by the stress-intensity factor, K. If plastic deformation is highly localized near the crack tip, then the plastic-zone size will also be completely controlled by K. Since acoustic emission is closely associated with plastic deformation, and the plastic deformation at a crack tip is controlled by K, it seems reasonable to expect that the acoustic emission will also be dependent on K.

A model to predict the relation between the acoustic emission from a cracked specimen and the stress-intensity factor for the crack was proposed by the authors.[18] It consists of assuming that the number of observed emission counts is proportional to the volume of the plastically deformed material at the crack tip, i.e.,

$$N = D'V_p \tag{3.1}$$

The plastic volume can be estimated from the size of the plastic zone, as given by McClintock and Irwin:[28]

$$r_y = \frac{1}{4\pi 2^{1/2}} \left(\frac{K}{\sigma_{ys}} \right)^2 \tag{3.2}$$

Assuming that the plastic zone is circular, V_p will be $B\pi r_y^2$ for a through crack in plate of thickness B. The following relation is then obtained:

$$N = D'V_p = D'B\pi r_y^2 = D'B\pi \frac{1}{32\pi^2} \frac{K^4}{\sigma_{ys}^4} = DK^4 \tag{3.3}$$

This relation shows that it is possible to relate the number of acoustic-emission counts for a given load directly to the stress-intensity factor for the crack at that load. This is a very important result, since it is actually the stress-intensity factor that controls the onset of rapid crack extension rather than the flaw size itself. Hence, if the critical stress-intensity factor, K_c, is known, acoustic emission can be used to directly determine how close the structure is to failure at a given load. Conventional nondestructive-testing techniques determine flaw sizes, but do not differentiate between those located in critical regions of high stress and the relatively harmless flaws located in lowly stressed regions.

The strong dependence of the acoustic emissions on the stress-intensity factor, as expressed in the above equation, indicates that, if more than one flaw is present, the largest flaw will be the one that contributes the majority of the acoustic emission, and smaller flaws will have a secondary effect.

The acoustic emission from embedded and surface flaws cannot be directly related to the stress-intensity factor, but reference must be made to the flaw size itself. The case of an embedded circular flaw (penny-shaped crack) will be worked out as an example; the results are easily generalized to more complex embedded and surface flaws. The only difference between this derivation and that for a through crack is that the plastic zone will be toroidal with a volume of $2\pi a\pi r_y^2$ (rather than $B\pi r_y^2$). This then leads to

$$N = D' V_p = D' 2\pi^2 a r_y^2 = D' 2\pi^2 a \frac{1}{32\pi^2} \frac{K^4}{\sigma_{ys}^4} = D'' a K^4 \qquad (3.4)$$

which shows that reference must now be made to the flaw size itself.

In actual situations, the exponent in the acoustic emission–stress intensity factor relation has been found to usually be different from 4; varying from 4 for some 7075-T6 specimens[29] to about 8 for beryllium.[18]   Figures 3.6 and 3.7 present acoustic-emission results from $1\frac{1}{2}$-in.-wide, 0.10-in.-thick single-edge-notch fracture-toughness specimens of N50A beryllium. References 18 and 30 present details of the specimen configuration and preparation. This figure shows

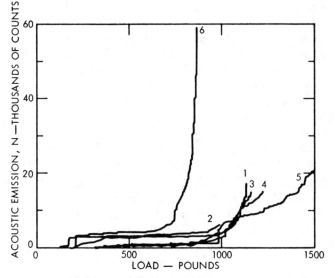

Fig. 3.6—Acoustic emission as a function of load for beryllium fracture-toughness specimens. (From Ref. 18)

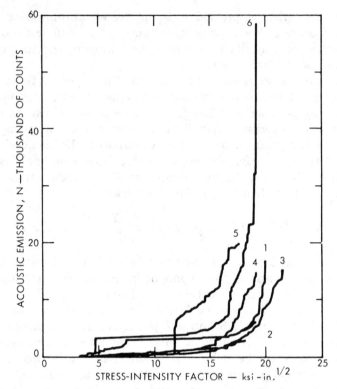

Fig. 3.7—Acoustic emission as a function of stress-intensity factor for the beryllium specimens of Fig. 3.6. (From Ref. 18)

that the stress-intensity factor rather than the load or nominal-stress level is indeed the parameter controlling the acoustic emission from these flawed specimens (when the amount of scatter often associated with acoustic-emission tests is taken into consideration). The exponent in the N–K relation was about 8 for these beryllium specimens.

Figures 3.8 and 3.9 show acoustic-emission results from four $1\frac{1}{2}$-in.-wide, 0.10-in.-thick single-edge-notch fracture-toughness specimens of 7075-T6 aluminum with fatigue cracks of various lengths. These results again show that the stress-intensity factor is the parameter controlling acoustic emission. The exponent for this material and thickness was 4—the value predicted from the model.

Tests were also conducted on 0.40-in.-thick, $2\frac{1}{4}$-in.-wide 7075-T6 specimens containing fatigue cracks in various locations. These tests were performed to see if the emission was altered by the presence of multiple flaws, and by the use of geometries other than a single-edge notch. The results of these tests are presented in Figs. 3.10 and 3.11, which show that K is again the controlling parameter, and the emission was not appreciably altered by the presence of multiple flaws.

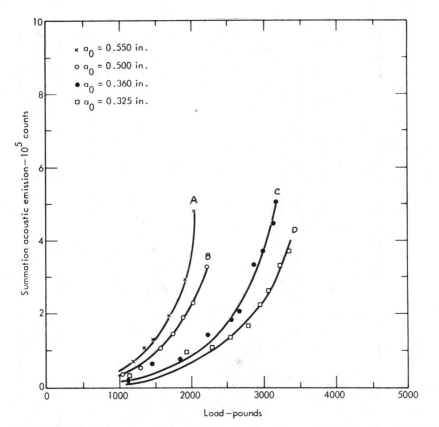

Fig. 3.8—Summation acoustic emission as a function of load for four SEN fracture-toughness specimens of 7075-T6 aluminum, with varying crack length. (From Ref. 29)

The exponent in the N–K relation was found to 5.4 for this material and thickness.

The disagreement between the theoretically predicted exponent of 4 and the results of actual tests could be caused by several factors. Some possible reasons for this discrepancy are the presence of twinning, such as would occur near the crack tip in beryllium, and microcrack formation and crack tunneling (pop-in) such as would occur in the aluminum. These factors would tend to give a higher exponent than that predicted, because they are factors that occur in addition to the plastic deformation considered in the model.

Investigators at Aerojet-General Corporation and other organizations have applied acoustic emission with multiple transducers, and triangulation techniques to the detection and location of flaws in large structures. They are able to locate flaws by measuring time delays between the arrival of acoustic emission pulses at various loca-

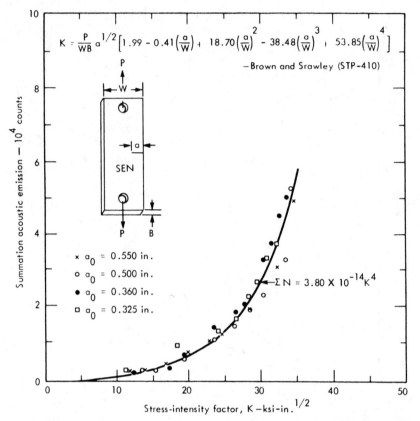

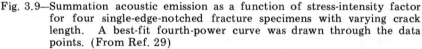

Fig. 3.9—Summation acoustic emission as a function of stress-intensity factor for four single-edge-notched fracture specimens with varying crack length. A best-fit fourth-power curve was drawn through the data points. (From Ref. 29)

tions. Figure 3.12 from Ref. 31 shows the fracture origin indicated by triangulation measurements, and the fracture origin determined from the fracture-surface markings in the failure of a 260-in.-diam rocket-motor case. This figure shows the good agreement between the location of the failure origin as determined by the two independent methods, and indicates the applicability of acoustic emission to the location of flaws in large structures.

3.5 USE OF ACOUSTIC EMISSION FOR DETECTION OF SUBCRITICAL FLAW GROWTH

The subcritical growth of flaws has recently become of considerable technological interest. The initial presence of small flaws in a structure often does not jeopardize a structure, since they are below

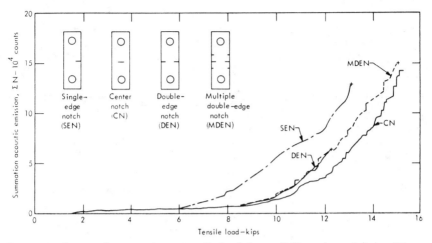

Fig. 3.10—Summation counts vs. tensile load for multiflawed specimens. (From
Ref. 29)

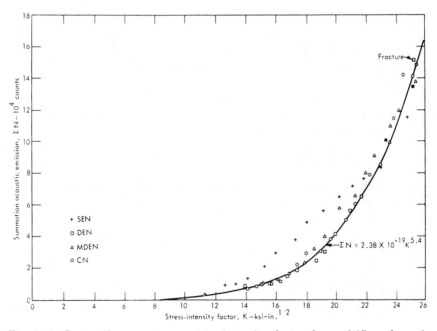

Fig. 3.11—Summation counts vs. stress-intensity factor for multiflawed specimens. (From Ref. 29)

the critical crack size. However, if these small flaws can grow, by a
process such as stress-corrosion cracking or fatigue-crack growth,
they may reach a critical size and cause a catastrophic failure of the
structure.

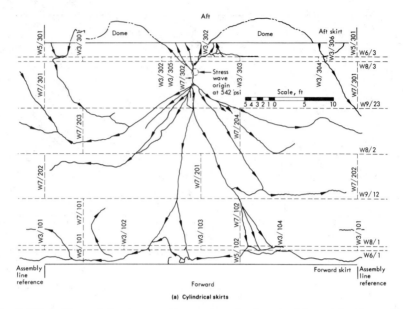

Fig. 3.12(a)—Map of fracture paths as viewed when looking at inside surface developed onto plane obtained by unrolling cylinder. Arrowheads indicate directions of fracture propagation.

Elastic stress waves and plastic deformation accompany subcritical crack growth, and acoustic emission has proved to be well suited to the detection of this growth.

Investigations of subcritical flaw growth using acoustic-emission techniques can be divided into four categories:

1. Detection of subcritical growth during rising load (pop-in).
2. Detection of crack growth at constant load, such as occurs during stress-corrosion cracking and hydrogen embrittlement.
3. Detection of crack growth due to a fluctuating load by continuous acoustic-emission monitoring.
4. Detection of crack growth due to a fluctuating load by intermittent overstressing.

Each of these detection techniques will now be discussed in detail.

Subcritical Growth during Rising Load

As has been mentioned earlier, one of the earliest uses of acoustic emission in fracture studies was the detection of pop-in during a fracture-toughness test.[14,15] Figure 3.13 shows the acoustic emission and notch-opening displacement measurements on a fatigue-cracked

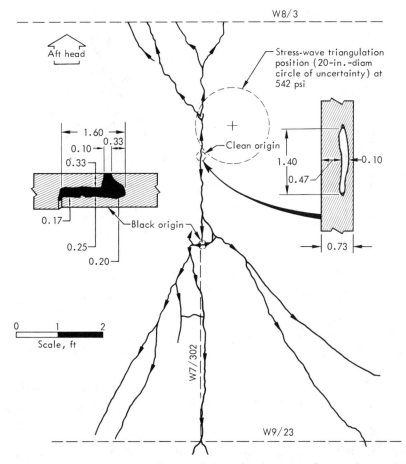

Fig. 3.12(b)—Detail in immediate vicinity of primary and secondary origins. Inset sketches show shapes and dimensions of the two origins. (From Ref. 31)

single-edge-notched fracture-toughness specimen of 7075-T6 aluminum. This figure shows that the beginning of acoustic emission is associated with the slight change in slope of the load–displacement curve that occurs at a load of about 5,000 lb. A rapid jump in displacement occurred at about 7,000 lb, which was accompanied by copius acoustic emission. The corresponding K_{Ic} value of 30.6 ksi-in.[1/2] agrees well with the results of numerous other investigators. This result shows that acoustic-emission techniques are well suited for the detection of pop-in.

Gerberich and Hartbower[32,33] at Aerojet-General have related the acoustic emission associated with pop-in to the new area de-

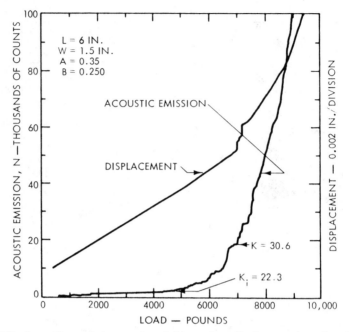

Fig. 3.13—Acoustic emission and notch-opening displacement as functions of
    load for a single-edge-notched specimen of 7075-T6 aluminum.
    (From Ref. 18)

veloped as a result of crack extension. They have found the following quantitative relationship for a wide variety of materials:

$$\Delta A \propto (\Sigma g)^2 E/K^2 \qquad (3.5)$$

When ΔA is the incremental area swept out by the crack, Σg is the
sum of the stress-wave amplitudes associated with that increment of
growth, E is the elastic modulus, and K the applied-stress-intensity
factor. This relation has also been found to hold for subcritical
crack growth due to stress-corrosion cracking. Thus, it is seen that
acoustic emission can be used to obtain a quantitative measurement
of crack growth during rising- and constant-load tests.

Continuous Monitoring of Fatigue-crack Growth

The fatigue-crack growth produced by fluctuating loads can be
detected by continuous monitoring of acoustic emission. Hartbower
et al.[34] have investigated the acoustic emission from D6AC steel,
with different heat treatments, during low-cycle fatigue. Their results are presented in Fig. 3.14, where Σg is the summation of the
amplitudes of the emission pulses. Figure 3.14 shows that it is possi-

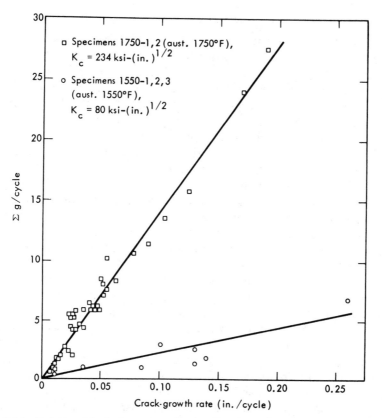

Fig. 3.14—Relationship between crack-growth rate and stress-wave emission for two conditions of D6AC steel. (From Ref. 34)

ble to detect the growth of fatigue cracks by continuous monitoring and, further, that the amounts of crack growth per cycle can be directly determined from the acoustic-emission data. However, the amount of emission for an increment of crack growth depends not only on the material but also on its previous history, such as its heat treatment. The results of Fig. 3.14 indicate that the amount of emission per cycle greatly decreases at low crack-growth rates, and that the crack-growth rate associated with high-cycle fatigue ($\sim 1 \mu$ in./ cycle) may be below the detection threshold of the instrumentation. Furthermore, it would often be inconvenient to continuously monitor a structure subjected to a large number of fatigue cycles and, in most cases, it would be more convenient to apply an intermittent overstressing with simultaneous monitoring for acoustic emission to the detection of the growth of fatigue cracks.

Intermittent Monitoring of Fatigue-crack Growth

As indicated in the previous section, it is possible to detect fatigue-crack growth by continuous acoustic-emission monitoring. However, the presence of background noise during service and the high expense would make continuous monitoring of acoustic emission impractical in most situations of technological interest. As an alternative to continuous monitoring, a procedure which takes advantage of the irreversibility of acoustic emission is possible.

If a cracked structure is loaded to a particular value of K and then unloaded, emission will not occur during reloading until this previous value of K is exceeded. It is possible to take advantage of this irreversible nature to determine whether or not a crack has grown during cyclic loading at a stress σ_w by periodically overstressing (proof testing) the structure at a stress σ_p ($> \sigma_w$) while monitoring for acoustic emission. If flaws have grown at σ_w since the previous overstress, then the stress-intensity factor during proof testing ($K \propto \sigma_p a^{1/2}$) will have increased, and emission will be observed during the proof test. Alternatively, if no crack growth occurred at σ_w, K_p would remain the same as during the previous proof test and no new plastic deformation (and, hence, no acoustic emission) would occur.

The applicability of this technique to the detection of fatigue-crack growth in fracture-toughness specimens of "trip" steel[35] and 7075-T6 aluminum[36] has been demonstrated. The results for these two materials were similar, and only the 7056-T6 data will be presented here.

Wedge-opening-loading (WOL)[37] specimens, as shown in Fig. 3.15, with $W = 2$, $B = 0.48$ were prepared from 7075-T6 plate stock. Three types of tests were conducted:

1. Rising-load fracture-toughness tests. These tests were conducted to determine K_{Ic}, the fracture toughness of the material, in the conventional manner. A fatigue crack was introduced into the specimen prior to the test, and acoustic emission was monitored during the test. The relationship between the acoustic-emission counts and stress-intensity factor was found to be

$$N = 2.46 \times 10^{-5} K^7 \qquad (3.6)$$

   where N is in counts, and K in ksi-in.$^{1/2}$. K_{Ic} was found to be 25.7 ksi-in.$^{1/2}$ (average of three tests).

2. Straight fatigue to failure test. These tests consisted of cycling the specimen between 50 and 800 lb, and measuring

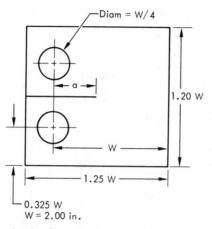

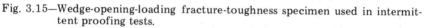

Fig. 3.15—Wedge-opening-loading fracture-toughness specimen used in intermittent proofing tests.

the crack length as a function of the number of fatigue cycles. No acoustic emission was taken during these tests. These tests provided the crack-growth data needed to determine the crack-growth-rate law for this material. The following relationship was found:

$$da/dn = 2.52 \times 10^{-8} \hat{K}^{2.69} \tag{3.7}$$

where a is in inches, and $\hat{K}$ in ksi-in.$^{1/2}$. This form of the crack-growth law was proposed by Paris[38] and agrees well with a large volume of experimental data.

3. Fatigue with intermittent proof. These tests consisted of introducing a 0.70-in. fatigue crack into the specimen by cycling to 800 lb. This was taken to be the point of zero fatigue cycles. The specimen was then cycled between 50 and 800 lb and proofed to 1,200 lb every 3,000 cycles. Acoustic emission was monitored during the proof loading, and the amount of acoustic emission observed (before it decreased to background) while holding at the proof load was noted. The results of these tests are presented in Table 3.1, and Figs. 3.16 and 3.17.

Since the acoustic emission is directly related to the stress-intensity factor, as also is the fatigue-crack-growth rate, it is possible to directly correlate acoustic emission, crack growth, and number of fatigue cycles. The N_t–n relation can be derived for any geometry for which the stress-intensity factor is known, and any material whose crack-growth law and N–K [eq. (3.6)] relation has been de-

Table 3.1—Results of Intermittent Proofing Test

No. of Fatigue Cycles (thousands of cycles)	Specimen 1			Specimen 2		
	Crack Length (in.)	Total Counts to 1.2 kips (counts)	Counts during Hold at 1.2 kips (counts)	Crack Length (in.)	Total Counts to 1.2 kips (counts)	Counts during Hold at 1.2 kips (counts)
0	0.700	360	0	0.700	305	0
3	0.719	165	0	0.725	158	0
6	0.746	429	0	0.749	121	0
9	0.781	454	0	0.771	185	0
12	0.815	750	0	0.792	83	0
15	0.841	353	0	0.810	115	0
18	0.878	720	0	0.830	491	0
21	0.919	911	0	0.855	304	0
24	0.965	934	0	0.885	429	0
27	1.028	1,040	1,000	0.909	488	0
30	1.100	33,780	15,300	0.941	600	12
33	1.206	158,000*	*	0.975	1,350	100
36	. . .	. . .	. . .	1.021	5,200	50
39	. . .	. . .	. . .	1.078	9,390	8
42	. . .	. . .	. . .	1.159	41,400	9,000
45	. . .	. . .	. . .	1.272	46,600*	*

*Broke before reaching maximum load during proof cycle.

termined. The special case of WOL specimen is of interest and will be worked out. The stress-intensity-factor relation for this geometry is[37]

$$K = \frac{Fa^{1/2}}{BW} G(a/W) = \frac{F}{B\sqrt{W}} \alpha^{1/2} G(\alpha) \tag{3.8}$$

where

$$G(\alpha) = 100(0.296 - 1.855\,\alpha + 6.557\,\alpha^2 - 10.17\,\alpha^3 + 6.389\,\alpha^4) \tag{3.9}$$

Taking eq. (3.7) in the form

$$da/dn = C\hat{K}^q \tag{3.10}$$

($C = 2.52 \times 10^{-8}$, $q = 2.69$ for the material under consideration) and combining it with the K relation leads to the following differential equation:

$$\frac{da}{dn} = W\frac{d(a/W)}{dn} = W\frac{d\alpha}{dn} = C\hat{K}^q = CK_w^q = C\left(\frac{F_w}{BW^{1/2}}\alpha^{1/2}G(\alpha)\right)^q \tag{3.11}$$

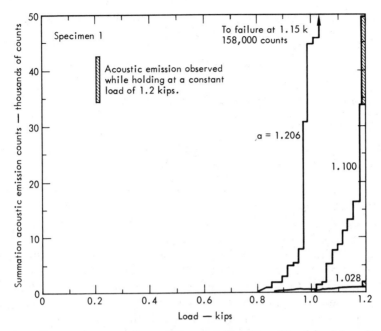

Fig. 3.16—Summation acoustic emission as a function of load for intermittently proofed cyclically loaded specimen. (From Ref. 36)

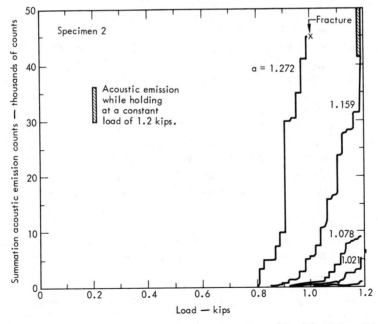

Fig. 3.17—Summation acoustic emission as a function of load for intermittently proofed cyclically loaded specimen. (From Ref. 36)

Solving the differential equation by separating variables and integrating leads to the following expression:

$$n(\alpha) = \eta(\alpha) \bigg/ \left[\frac{C}{W} \left(\frac{F_w}{BW^{1/2}} \right)^q \right]$$

(3.12)

where $\eta(\alpha)$ is the following integral, which must be evaluated numerically

$$\eta(\alpha) = \int_{\alpha_o}^{\alpha} \frac{dx}{[x^{1/2} G(x)]^q}$$

(3.13)

The value of N_t corresponding to the crack length $a = \alpha W$ and number of fatigue cycles n can be calculated by determining K for the known crack length, and proof load [eq. (3.8)], and substituting the result into the N-K relation [eq. (3.6)].

The results of such calculations for the 7075-T6 aluminum are presented in Fig. 3.18, along with the experimental results of Table 3.1. This figure shows the good agreement between the theoretical and experimental results, which indicates that the model provides an adequate tool for the analysis of these types of tests. The emission observed while holding at the proof load provides another piece of information regarding the imminence of failure which, in addition, can be used to determine the optimum number of cycles between proofing. This is dealt with in detail in Refs. 35 and 36. It is worth noting that the results of these tests indicate that the intermittent overstressing of these specimens did not adversely affect the fatigue-crack-propagation behavior and, thus, the proofing did not shorten their fatigue life.

Work is presently under way to demonstrate the applicability of this technique to welded pressure vessels,* both with and without known initial flaws. The theoretical and experimental results for the 7075-T6 WOL specimens presented above both show that intermittent proofing provides early warning of impending failure due to the growth of fatigue cracks, and that this technique should be highly useful in assessing the integrity of actual engineering structures.

Stress-corrosion Studies

The critical stress-intensity factor can be drastically lowered when a specimen under load, containing a sharp crack, is subjected to

---

*This work has been completed and is reported in "Verification of Structural Integrity of Pressure Vessels by Acoustic Emission and Periodic Proof Testing," *Testing for Prediction of Material Performance in Structures and Components*, ASTM STP 515, 158–170 (1972).

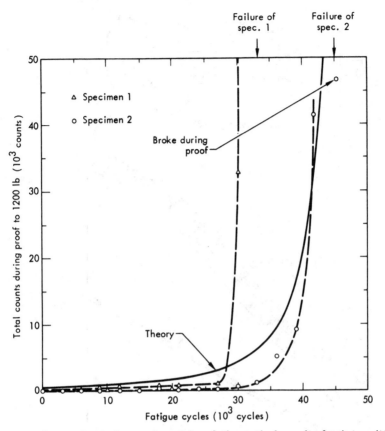

Fig. 3.18—Comparison of experimental and theoretical results for intermittent proof tests. (From Ref. 36)

hostile environments.[39] The mechanism for this deterioration in strength is in some cases quite complex, but the general term used to describe this phenomenon is stress corrosion. The accepted designation for the plane-strain fracture toughness under these environments is K_{ISCC}[39] or the stress-corrosion fracture toughness. It can have values ranging from approximately 25 percent to 90 percent of its value under normal conditions, depending on the alloy and environment.

The linear-compliance specimen was chosen for this study (see Ref. 40 for detailed information concerning this specimen). It has the feature that for a given load, P, the stress-intensity factor is independent of the crack length over several inches. Figure 3.19 is a drawing of this specimen and Fig. 3.20 shows the specimen in a specially designed loading fixture. The strain-gaged bolt shown as part

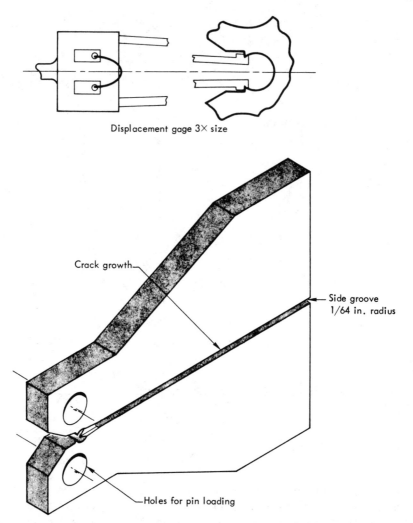

Displacement gage 3× size

Fig. 3.19—Linear-compliance specimen used in fracture-toughness tests, showing
side groove and notch geometry used for displacement gage.

of the fixture in Fig. 3.20 was used to load the specimen and also to
act as a load cell. The expression for the stress-intensity factor for
the geometry chosen is

$$K = 10F \qquad (3.14)$$

We see that no crack length is contained in this expression and
the stress-intensity factor, K, is simply proportional to the load, F.
This expression is only valid over the portion of the specimen con-

Fig. 3.20—Linear-compliance specimen in special loading fixture used in stress-corrosion studies.

taining the curved section. This direct proportionality between load and stress intensity allows a plot of the stress-intensity factors to be made directly from the output of the load cell.

It has been observed on some uranium alloys that slow crack extension will occur at stress-intensity values of approximately 85 ksi-in.$^{1/2}$. These specimens were loaded to this value with the bolt while recording the stress-intensity factor and acoustic-emission data as a function of time on X-Y plotter. The drop-off in load (stress intensity) and acoustic emission was then plotted as crack extension occurred. The stress-intensity factor corresponding to the arrest of the crack is K_{ISCC}, as mentioned previously.

Figure 3.21 is the result of a stress-corrosion test conducted on a uranium-0.3-percent titanium specimen. The acoustic emission is observed to begin at a stress intensity of 30 ksi-in.$^{1/2}$. At a value of 90 ksi-in.$^{1/2}$, the load bolt on the stress-corrosion fixture was locked in place. Some slow crack extension continued for a period of 15 or 20 min at reduced rate. The acoustic emission also shows a decreasing rate, following the drop in stress-intensity factor. At this time, a few drops of 3 percent salt solution were placed in the side groove of the specimen. The acoustic emission immediately began to increase at a high rate and, shortly thereafter, the load on the bolt began to decrease. The crack growth due to this corrosion process was visually evident. After approximately 15 min, the crack began to arrest and the acoustic emission approached an asymptotic value.

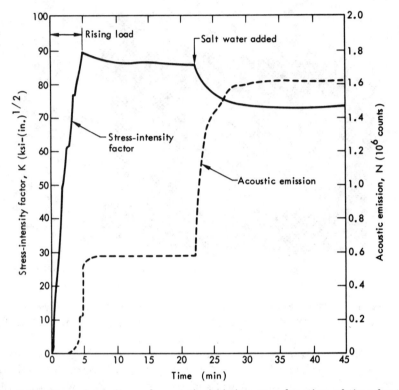

Fig. 3.21—Stress intensity and acoustic emission as a function of time for the fixed-grip condition of a uranium–0.3-percent titanium linear-compliance fracture-toughness specimen undergoing stress corrosion due to 3-percent salt solution.

This specimen was later taken to failure and the failure occurred by arm break-off; hence, it is indefinite at this time whether or not the arrest K is a legitimate K_{ISCC}. Later tests on wrought uranium tend to show that this was not a legitimate arrest, and that the specimen probably arrested out of the plane of the side groove.

Figure 3.22 is a similar test on the same alloy, with the exception that this specimen contained an electron-beam weld in the plane of the side groove. This specimen exhibited slow crack growth at stress intensities lower than the specimen without the weld. Distilled water and methyl alcohol were added to the crack surface with little short-term effect on the load or acoustic emission. The addition of salt water did cause an immediate reaction that quickly sent the acoustic-emission counts off scale. After a short delay, the load began to drop off. This time a region of stable crack extension was noticed over a period of approximately 10 min. By this time, the crack had ex-

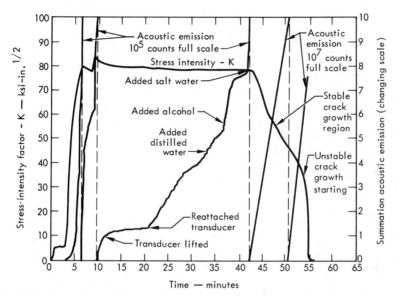

Fig. 3.22—Acoustic emission and stress-intensity factor as a function of time for a uranium–0.3-percent titanium linear-compliance fracture specimen undergoing stress corrosion in a fixed-grip condition. Specimen contained an electron-beam weld in the plane of the crack.

tended almost to the end of the curved region (see Fig. 3.19) and then became unstable, resulting in rapid crack propagation and failure.

Hydrogen Embrittlement

Acoustic-emission techniques have been used to determine the rate of cracking of high-strength steel due to hydrogen embrittlement, making it possible to predict failure in certain structural components undergoing hydrogen attack.

The specimen used (Fig. 3.19) was the same as in the stress-corrosion studies, with the exception that the tests were performed using a dead-weight load rather than the constant-displacement fixture. The specimens were made from 4340 steel in the 220-Ksi yield-strength range. They were cathodically charged with hydrogen, cadmium plated and baked for 3 hr at 150°C, fatigue cycled to grow a crack into the side groove, and placed in a dead-weight-load frame. An acoustic-emission transducer was taped to the side, and a COD gage was inserted to measure crack-opening displacement as the crack grew. The summation of acoustic-emission counts and crack-opening displacement was then recorded as a function of time for several

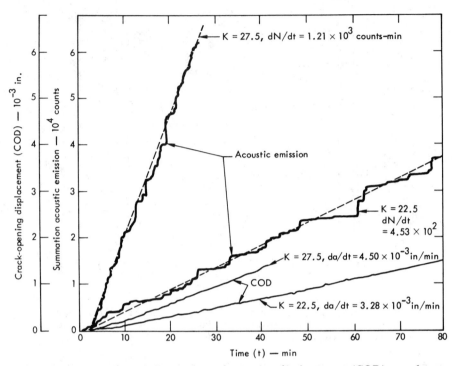

Fig. 3.23—Acoustic emission and crack-opening displacement (COD) as a function of time for a linear-compliance specimen of 4340 undergoing crack growth due to hydrogen embrittlement. Data were obtained for two values of dead-weight load, thus giving two values of stress intensity K. (From Ref. 41)

dead-weight-load conditions. This amounts to recording at different values of the stress-intensity factor K, because of the proportional relationship between load and K, as mentioned previously.

Figure 3.23 shows the summation of acoustic-emission signals and crack-opening displacement from one of the specimens. Data were first recorded at a load that would give a stress-intensity factor of 22.5 ksi-in.$^{1/2}$. After approximately 0.250 in. of crack growth, the load was increased so that K had a value of 27.5 ksi-in.$^{1/2}$. Note that the acoustic emission and crack-opening displacements are linear functions of time and the slopes of both curves are strongly dependent on K. The rate of change of cracked area $d(\Delta A)/dt$ was obtained by multiplying the linear crack-growth rate $d(\Delta a)/dt$ by the net section thickness, B. The crack-growth rate was calculated from the slope of the COD-time curve and the compliance vs. crack-length curve. The data for several values of K showed that the rate of crack growth varies linearly with K.

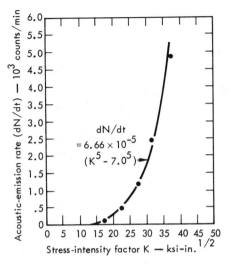

Fig. 3.24—Acoustic-emission rate as a function of stress-intensity factor for a hydrogen charged, linear-compliance fracture specimen of 4340 steel. (From Ref. 41)

The slopes of the summation emission on curves were taken for several values of K and are shown plotted in Fig. 3.24. A best fit of these data gives

$$dN/dt = 6.66 \times 10^{-5}\ (K^5 - K_o^5) \tag{3.15}$$

where K_o = 7.0 ksi-in.$^{1/2}$ is the threshold stress-intensity factor below which no cracking occurred.

The plane-strain fracture toughness for this material was 50 ksi-in.$^{1/2}$. Substitution of this value for K in eq. (3.15) results in a critical acoustic-emission rate of 2×10^4 counts/min. Several specimens in the form of cathodically charged, cadmium-plated and loaded bolts were acoustically monitored. In all cases, failure occurred at acoustic-emission rates within ±15 percent of the predicted value.

These results suggest that continuous acoustic-emission monitoring of certain structures containing an advancing hydrogen crack will allow prediction of impending failure.

A more detailed description of this work can be found in Ref. 41. The most important conclusion reached in this study is that, regardless of the initial loading, nucleation time for cracking to begin, or any other factors connected with geometrical differences in specimens, the acoustic-emission rate can be used to determine the time for and stress-intensity factor at the onset of rapid fracture. Since brittle fracture is completely controlled by the stress-intensity factor, the acoustic-emission rate can be used as the criterion to predict failure.

3.6 SUMMARY AND CONCLUSIONS

The results of investigations summarized in this chapter indicate the wide applicability of acoustic emission to experimental studies in fracture mechanics. A brief historical review of acoustic-emission studies on both rocks and metals is provided, and a discussion of the suitable transducers and electronics for acoustic-emission studies is presented. Acoustic emission is the characteristic and irreversible sound emitted by a material when it is deformed. Hence, acoustic emission is well suited to studies in which plastic deformation occurs. The onset of plastic deformation in materials is controlled by the presence of stress raisers, such as cracks and other types of flaws— hence, acoustic emission provides a useful tool for flaw detection. This aspect is treated in some detail, and it is concluded that acoustic-emission techniques are well suited to the detection of the presence of flaws in structures. In some instances, the stress-intensity factor of a crack for a given load can be directly determined using acoustic emission, and triangulation techniques can be used to locate flaws in large structures.

Subcritical flaw growth, such as that which occurs during pop-in, stress-corrosion cracking, fatigue, and hydrogen embrittlement, results in the release of elastic strain energy and plastic deformation. These processes give rise to acoustic emission, and the uses of acoustic emission to the detection of subcritical flaw growth are indicated. The results of investigations summarized in this chapter indicate that acoustic-emission techniques are well suited to the detection of the growth of subcritical flaws and, furthermore, that quantitative information regarding the imminence of failure can often be obtained from acoustic-emission data.

In conclusion, acoustic emission appears to be a very useful tool for the detection of the presence and growth of cracks and flaws in laboratory specimens and engineering structures.

REFERENCES

1. Hodgson, E. A., *Dominion Observatory Rockburst Research 1938–1945*, Department of Mines and Technical Surveys, Canada, Dominion Observatories, 20 (1958).
2. Obert, L., and Duvall, W., "Seismic Methods of Detecting and Delineating Sub-Surface Subsidence," U.S. Bureau of Mines Report of Investigation No. 5882 (1961).

3. Persson, T., and Hall, B., "Micro-Seismic Measurements for Predicting the Risk of Rock Failure and the Need for Reinforcement in Underground Cavities," Publications of the Royal Institute of Technology, Stockholm (1957).

4. Antsyferov, M. S., ed., *Seismo-Acoustic Methods of Mining*, translated from Russian by S. E. Hall, Consultants Bureau, New York (1966).

5. Knill, J. L., Franklin, J. A., and Malone, A. W., "A Study of Acoustic Emission from Stressed Rock," *Intl. Jnl. of Rock Mech. and Mineral Sci.*, **5**, 87-121 (1968).

6. Scholz, C. H., "The Frequency-Magnitude Relation of Microfracturing in Rock and Its Relation to Earthquakes," *Bull. of the Seismological Soc. of Amer.*, **58** (1), 399-415 (Feb. 1968).

7. Scholz, C. H., "Microfracturing and the Inelastic Deformation of Rock in Compression," *Jnl. of Geophysical Res.*, **73** (4), 1417-1432 (Feb. 1968).

8. Scholz, C. H., "Experimental Study of the Fracturing Process in Brittle Rock," *Jnl. of Geophysical Res.*, **73** (4), 1447-1454 (Feb. 1968).

9. Schofield, B. H., "Acoustic Emission under Applied Stress," Aeronautical Research Laboratory Report No. ARL-150 (1961), available from Office of Technical Services, U.S. Department of Commerce, Washington, D.C.

10. Kaiser, J., "Untersuchungen uber das auftreten Geraushen beim Zugversuch," PhD thesis, Techn. Hochsch., Munich, 1950; see also *Arkiv fur das Eisenhuttenwesen*, **24**, 43-45 (1953).

11. Tatro, C. A., and Liptai, R., "Acoustic Emission from Crystalline Substances," *Proc. of the Symposium on Physics and Nondestructive Testing*, Southwest Research Institute, San Antonio, Tex., 145-158 (1962).

12. Kerawalla, J. L., "An Investigation of the Acoustic Emission from Commercial Ferrous Materials Subjected to Cyclic Tensile Loading," PhD thesis, University of Michigan, Ann Arbor (1965).

13. Mitchell, L. D., "An Investigation of the Correlation of the Acoustic Emission Phenomenon with the Scatter in Fatigue Data," PhD thesis, University of Michigan, Ann Arbor (1965).

14. Romine, H. E., "Determination of the Driving Force for Crack Initiation from Acoustic Records of G_c Tests on High Strength Materials for Rocket Motor Cases," Naval Weapons Laboratory Report NWL 1779 (Oct. 1961).

15. Jones, M. H., and Brown, W. F., "Acoustic Detection of Crack Initiation in Sharply Notched Specimens," *Matls. Res. and Stds.*, **4**, 120-129 (1964).

16. Green, A. T., Hartbower, C. E., and Lockman, C. S., "Feasibility Study of Acoustic Depressurization System," Aerojet General Corporation Report No. NAS 7-310 (1965).

17. Schofield, B. H., "Investigation of Applicability of Acoustic Emission," Air Force Matls. Lab. Tech. Rept. No. AFML-TR-65-106 (1965).

18. Dunegan, H. L., Harris, D. O., and Tatro, C. A., "Fracture Analysis by Use of Acoustic Emission, *Engrg. Fract. Mech.*, **1** (1), 105-122 (June 1968).

19. Dunegan, H. L., Tatro, C. A., and Harris, D. O., "Acoustic Emission Research," Lawrence Radiation Lab. Rept. No. UCID-4868, Rev. 1 (1964).

20. Cook, N. G. W., "The Seismic Location of Rockbursts," *Proc. of the Fifth*

Symposium on Rock Mechanics, University of Minnesota, Minneapolis, 493–516 (1962).

21. Hutton, P. H., "Detection of Incipient Failure in Nuclear Reactor Pressure Systems Using Acoustic Emission," Battelle Northwest Rept. No. BNWL-997 (Apr. 1969).

22. Green, A. T., "Stress Wave Emission and Fracture of Prestressed Concrete Reactor Vessel Materials," Matls. Tech. AGC Rept. No. 4190 (June 1969).

23. Engle, R. B., "Acoustic Emission and Related Displacements in Lithium Fluoride Single Crystals," PhD thesis, Michigan State University, East Lansing (1966).

24. Fisher, R. M., and Lally, J. S., "Microplasticity Detected by an Acoustic Technique," *Canadian Jnl. of Physics*, 45, 1147–1159 (1967).

25. Dunegan, H. L., and Tatro, C. A., "Acoustic Emission Effects During Mechanical Deformation," *Techniques of Metals Res.*, 5, pt. 2, pp. 273–312 (1971) (editor, R. Bunshah), John Wiley & Sons.

26. Gilman, J. J., "Progress in Microdynamical Theory of Dislocations," *Proc. of the 5th U.S. Natl. Cong. of Appl. Mech.*, Amer. Soc. of Mech. Engrs., New York (1966).

27. Dunegan, H. L., and Tatro, C. A., "Passive Pressure Transducer Utilizing Acoustic Emission," *Rev. of Scient. Instr.*, 38 (8), 1145–1147 (Aug. 1967).

28. McClintock, F. A., and Irwin, G. R., "Plasticity Aspects of Fracture Mechanics," *Fracture Toughness Testing and Its Applications*, ASTM Special Technical Publication No. 381, Amer. Soc. for Test. and Matls., Philadelphia, 84–113 (1965).

29. Dunegan, H., and Harris, D., "Acoustic Emission—A New Nondestructive Testing Tool," *Ultrasonics*, 7 (3), 160–166 (July 1969).

30. Harris, D. O., and Dunegan, H. L., "Fracture Toughness of Beryllium," *Jnl. of Matls.*, 3 (1), 59–72 (Mar. 1968).

31. Srawley, J. E., and Esgar, J. B., "Investigation of Hydrotest Failure of Thiokol Chemical Corporation 260-inch-diameter SL-1 Motor Case," NASA Technical Memorandum NASA TMX-1194 (Jan. 1966).

32. Gerberich, W. W., and Hartbower, C. E., "Some Observations on Stress Wave Emission as a Measure of Crack Growth," *Intl. Jnl. of Fract. Mech.*, 3 (3), 185–192 (Sept. 1967).

33. Hartbower, C. E., Gerberich, W. W., and Liebowitz, H., "Investigation of Crack-Growth Stress-Wave Relationships," *Engrg. Fract. Mech.*, 1 (2), 291–308 (Aug. 1968).

34. Hartbower, C. E., Gerberich, W. W., and Crimmins, P. P., "Monitoring Subcritical Crack Growth by Detection of Elastic Stress Waves," *Weld. Jnl.*, 47 (1), 1s–18s (Jan. 1968).

35. Dunegan, H. L., Harris, D. O., and Tetelman, A. S., "Detection of Fatigue Crack Growth by Acoustic Emission Techniques," *Materials Evaluation*, 28 (10), 221–227 (Oct. 1970).

36. Harris, D. O., Dunegan, H. L., and Tetelman, A. S., "Prediction of Fatigue Lifetime by Combined Fracture Mechanics and Acoustic Emission Techniques," *Proc. of the Air Force Conference on Fatigue and Fracture of Aircraft Structures and Materials*, Air Force Flight Dynamics Laboratory Report AFFOL TR 70-144, pp. 459–471 (1970).

37. Wessel, E. T., "State of the Art of the WOL Specimen for K_{Ic} Testing," *Intl. Jnl. of Engrg. Fract. Mech.*, **1** (1), 77–103 (June 1968).
38. Paris, P. C., "The Fracture Mechanics Approach to Fatigue," *Proc. of the Tenth Sagamore Conference*, Syracuse University Press, Syracuse, New York, 107 (1965).
39. Johnson, H. H., and Paris, P. C., "Sub-critical Flaw Growth," *Intl. Jnl. of Engrg. Fract. Mech.*, **1** (1), 3–45 (June 1969).
40. Mostovoy, S., Crosley, P. B., and Ripling, E. J., "Use of Crack-Line-Loaded Specimens for Measuring Plane-Strain Fracture Toughness," *Jnl. of Matls.*, **2** (3), 661–681 (Sept. 1967).
41. Dunegan, H. L., and Tetelman, A. S., "Non-destructive Characterization of Hydrogen-embrittlement Cracking by Acoustic Emission Techniques," *Intl. Jnl. of Engrg. Fract. Mech.* **2** (4), 387–402 (June 1971).

4

COMPLIANCE MEASUREMENTS

by

R. T. Bubsey, D. M. Fisher,
M. H. Jones, and J. E. Srawley
National Aeronautics and
Space Administration
Lewis Research Center
Cleveland, Ohio

4.1 SYMBOLS

A = projected crack area
a = crack length
B = specimen thickness
C = specimen compliance (e/F)
D = major diameter of notched round bar
d = minor diameter of notched round bar
E = Young's modulus
e = displacement corresponding to F at point of application of F
F = load applied to specimen
$\mathcal{G}$ = fixed-grip strain-energy-release rate with crack extension or crack-extension force
$\mathcal{G}_I$ = G for opening mode of crack extension
H = depth of split arm
H_a = depth of tapered split arm at crack edge
H_F = depth of tapered split arm at load line
K = stress-intensity factor of elastic-stress field in vicinity of crack edge
K_{Ic} = plane-strain fracture toughness or critical value of K in opening mode
L = length of bending moment arm
M = applied bending moment
S = support span
W = specimen width
σ_N = average net-section stress

4.2 INTRODUCTION

Compliance is the amount of displacement per unit of applied force, the reciprocal of stiffness. A compliance coefficient is a particular kind of flexibility-influence coefficient (as used in texts on engineering mechanics), a ratio of displacement to force that represents the linearly elastic, static response of a body to the action of a pair of equal and opposite colinear forces. In practice, these opposite forces are the resultants of localized distributions of contact pressure (not necessarily homologous), and it is usual to refer to only one of the pair as the load and to omit mention of the other (the reaction). For example, the forces are applied to the simple specimen shown in Fig. 4.1 by the pressure of loading pins on the surfaces of the holes. A compliance coefficient might also be defined for a pair of couples which bend or twist a body, but this reduces for small deformations to the compliance coefficient for the colinear resultants of two pairs of equal and opposite forces.

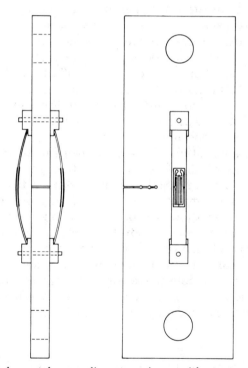

Fig. 4.1—Single-edge-notch compliance specimen with strut gages and stepped blocks.

4.3 APPLICATION TO LINEAR-FRACTURE MECHANICS

When a body contains a crack, the compliance for a given position and direction of the load will depend on the dimensions of the crack and, in certain simple cases, the derivative of the compliance with respect to crack area can be taken to be equal to $2\mathcal{G}/F^2$, where $\mathcal{G}$ is the fixed-grip strain-energy-release rate with crack extension,[1] and F is the magnitude of the load. The stress-intensity factor, K, can be obtained from known relations between components of $\mathcal{G}$ and K. These symbols, and the shorter name for $\mathcal{G}$, crack-extension force, were introduced by Irwin,[2] who later elaborated the concept to distinguish the three components of $\mathcal{G}$ (and of K) that correspond to the three possible orthogonal modes of crack extension.[3] These modes relate to the three orthogonal components of the relative displacement of corresponding points on the two opposite surfaces of a crack near the crack front. For a plane crack, the relative displacements are: in mode I, normal to the crack plane; in mode II, normal to the crack front in the crack plane; mode III, parallel to the front*. The compliance-measurement method can be applied in a simple manner only to plane cracks. It seems to have been applied only to obtain $\mathcal{G}_I$, the first or opening-mode $\mathcal{G}$ value, though it could be applied to either of the shear modes, which are of less practical interest. It is sufficient, however, to discuss the application of the method to $\mathcal{G}_I$, since the adaptation to the other components is straightforward once the general principles and experimental precautions are understood. The subscript I is omitted when it is clear from the context that only mode I is considered.

The useful application of the compliance method is, in fact, restricted to cracks of simple shape, such that either $\mathcal{G}$ or K can be taken as independent of position along the crack front or the values of $\mathcal{G}$ for individual segments of the leading edge of the crack are not significant. The method has been applied to the concentrically cracked round bar in axial tension by Lubahn,[4] but the precision of results obtained with this very stiff form of specimen is much inferior to the precision of available mathematical methods.[5] Nevertheless, there remains considerable scope for use of the compliance method because there is a variety of more compliant through-cracked plate specimens which might find applications for special purposes (see, for example, Ref. 6).

---

*For a segment of crack front which is a general space curve, the directions of relative displacement correspond to the unit vectors of differential geometry; the binormal for mode I, the principal normal for mode II, and the tangent for mode III.

The particular advantage of the compliance-measurement method is that the loading fixtures used can be identical to those needed for fracture tests or crack-propagation tests of materials. (However, only one size of prototype specimen is needed for the compliance measurements and, if a range of specimen sizes is needed for materials tests, the fixtures should be carefully scaled to the specimen.) If, as an alternative to compliance measurements, a numerical method of analysis is used to obtain K for a practical specimen, the boundary conditions which represent the load often have to be simplified to the point where they may introduce appreciable error. On the other hand, numerical methods are free from the sources of experimental error that can affect $\mathcal{G}$ values obtained from compliance measurements. In certain cases, like the concentrically cracked round bar, numerical methods can give much more accurate results than the compliance method. Whenever possible, both methods should be used so that the results can be compared and the sources of differences between them considered. As a general rule, the measurement precision needed to obtain good results by the compliance method is no less than that needed for Young's modulus (USA Standards B101.1-1963 and ASTM Standard E-111-61). This high precision is necessary because $\mathcal{G}$ is obtained by differentiation, which produces a relative error in $\mathcal{G}$ several times as large as that in the primary data.

4.4 RELATION BETWEEN $\mathcal{G}_I$ AND K_I

There is a minor degree of ambiguity about the relation between values of $\mathcal{G}_I$ obtained from experimental compliance measurements and the corresponding values of K_I obtained by numerical analysis of a two-dimensional analog of the compliance-calibration specimen. Because of this ambiguity (which is discussed below), the results of compliance measurements presented in this chapter are given in terms of $\mathcal{G}_I$, while the comparative results obtained by numerical analysis are given in terms of K_I. All results are given in the form of dimensionless quantities which are directly comparable when Poisson's ratio is zero.

The results of two-dimensional numerical analysis can apply either to a state of plane stress throughout the body, in which case $\mathcal{G}_I = K_I^2/E$, or to a state of plane strain throughout the body, in which case $\mathcal{G}_I = K_I^2/E(1 - \nu^2)$, where E is Young's modulus and ν is Poisson's ratio. The state of deformation is not uniform throughout a compliance-calibration specimen, but the appropriate relation between $\mathcal{G}_I$ and K_I should refer only to the region in the immediate vicinity of the front of the simulated crack. The reason for this localization is that K_I is defined as the limit of $\sigma_y(2\pi r)^{1/2}$ as r approaches zero in the direction $\theta = 0$, where r and θ are polar coordinates referred to the crack tip (see Chapter 2). Consequently, the relation between $\mathcal{G}_I$ and K_I should refer to the limit of the state of deformation under the same condition. However, since the compliance-calibration specimen has a slot (or notch) of finite-tip radius in place of a crack, the localization can be considered uncertain to this extent.

In the case of a plate specimen with a through-thickness slot, the state of deformation varies in the thickness direction as well as with distance from the slot front. The value of $\mathcal{G}_I$ from compliance measurements is an average value over the thickness, and the relation of K_I to the average value of $\mathcal{G}_I$ may be assumed to lie between the aforementioned relations for uniform plane stress and uniform plane strain. If the slot is narrow compared to the thickness (as it should be in a compliance-calibration specimen where the intent is to simulate a crack of negligible width), the average degree of constraint will approach plane strain in the immediate vicinity of the slot front, but the extent of the deviation from strict plane strain will depend on the ratio of specimen thickness to slot length. In the absence of a reliable three-dimensional analysis of the specimen, the relation between $\mathcal{G}_I$ and K_I is uncertain within the bounds of plane stress and plane strain.

It is worth noting also that there is a distinction to be made between a slotted plate specimen used for compliance measurements and a similar, but cracked, plate specimen used in a materials test, where there is a plastic zone around the crack front. In a materials test, the ratio of the size of the plastic zone to the specimen thickness is considered to determine the degree of constraint around the front of a propagating crack. For example, in a fatigue-crack-propagation experiment on a thin plate, the average constraint is considered to approach plane strain when the plastic zone is small compared with the thickness, and to approach generalized plane stress when the plastic zone is large compared to the thickness. In a compliance-calibration specimen, there should be no plastic zone; the corresponding significant dimension is the slot-tip radius.

4.5 OUTLINE OF PROCEDURE

A prototype is prepared of the specimens that are expected to be used for fracture tests or crack-propagation tests on materials. The material for this prototype need not be the same as the materials to be tested, but it should be homogeneous, isotropic, and should have a wide range of linear elastic behavior (up to a uniform tensile strain of at least 0.003). The Young's modulus should be accurately known or determined. The lateral dimensions should be strictly proportioned to the intended test specimens in all respects except for the simulated crack, which is represented by a narrow machined slot, initially shorter than the shortest crack length to be used in a test specimen. The specimen thickness must be uniform, and at least ten times the slot width.

A set of compliance measurements is made on the prototype, and the slot is extended by a small increment between each pair of consecutive measurements. The slot length for each increment must be measured with the greatest possible accuracy. The procedure is repeated until the slot length is greater than the longest crack to be used in a test specimen. The compliance coefficients are expressed in the dimensionless form EBe/F, where E is Young's modulus, B is the specimen thickness, and e/F is the displacement per unit load. This set of results is then fitted to a polynomial in a/W, where a is the slot length and W is the specimen width by standard statistical methods of "best fit." Choice of type and degree of polynomial function, and possibility of using transformed variables, are matters which call for some statistical sophistication and may require the advice of specialists. The fitting function is differentiated with respect to a/W to obtain a derived polynomial in a/W which represents $d(EBe/F)/d(a/W)$, which is equal to $2E\mathcal{G}B^2W/F^2$, as shown in Refs. 1, 3, and 7. This dimensionless function then applies to a specimen of any size that is geometrically similar to the prototype, and of any material which has a well-defined Young's modulus.

4.6 PRACTICAL EXAMPLE

Specimen Description

The single-edge-notch tension-compliance calibration of Srawley, Jones, and Gross[8] is representative of the precision and accuracy necessary for the development of a relationship sufficient for experimental $\mathcal{G}$ determinations. Compliances were determined for $\frac{1}{2}$-in.-thick, 3-in.-wide, centrally loaded 7075-T6 aluminum specimens over a crack length-to-specimen width (a/W) range of 0 to 0.5, as shown in Fig. 4.1. Displacement measurements were made of central sections of the specimens to reduce possible errors associated with the large strains occurring near the loading pins. Preliminary studies of photoelastic and compliance specimens were made to determine the distance from the central crack necessary to avoid its stress-field influence. Based on these studies, a minimum gage length of twice the specimen width was chosen. Compliance determinations were established for two specimen lengths. A single 24-in.-long specimen was calibrated using gage lengths of 6, 8, and 10 in. Duplicate 12-in.-long specimens were calibrated with an 8-in. gage length. The 12-in. specimen was that estimated for optimum material economy based on the length-to-width ratio.

Compliance Transducer

Displacements were measured with specially constructed flexed-strut gages, shown in Fig. 4.1. These gages were made of a 0.060-in.-thick β 120 titanium alloy. Epoxy-backed resistance strain gages were bonded to each strut centrally located in respect to width and length. The strut gages were held on both faces of the specimen in pairs of stepped blocks which were pinned to the specimen at the various gage lengths. By using stepped blocks of different lengths, the same strut gages were used for all gage lengths. When the gages were not in use, they were flexed in keeper blocks to minimize zero drift due to creep of the gage-mounting cement. The displacement measurements had an expected accuracy of $\pm\frac{1}{2}$ percent. A commercial strain indicator was used as the power supply and output indicator for the strut gages. The relationship between strain and chord length of such a flexed strut is nonlinear and is expressible in terms of elliptical integrals, if end effects are neglected. The strut gages were calibrated directly with a super micrometer to an estimated precision of ± 0.00003 in.

Experimental Procedure

To provide the greatest accuracy in determining the load, the testing-machine load ranges were chosen so that the maximum load for each compliance measurement was a major fraction of the load range. The specimen-loading train included at each specimen end tandem pin joints intended to allow the specimen to bend freely. Any effect of friction in these pin joints was neglected, however.

Specimen compliances were determined for 11 crack-to-width ratios over a range 0 to 0.5. Cracks were simulated by a 0.010-in.-wide jewelers saw cut terminating at a $\frac{1}{16}$-in.-diam drilled hole.† Prior to a compliance-data run, specimens were twice taken to the maximum load established for the particular compliance determination to establish practically linear response of displacement to load.

---

†The crack length, a, was taken to be somewhat less than the slot length. It is now felt that a should be equal to the slot length.

Individual strut gages were mounted on both sides of the specimen (Fig. 4.1) and separate load–displacement determinations made for each gage at a particular crack length. Averaging of the load-displacement curves compensated for errors due to out-of-plane bending. However, in cases where the two curves differed by more than 2 percent, the loading train was adjusted and the trial rerun.

Calculations

Primary data reduction was performed by fitting the compliance values to polynomials in a/W using a least-squares-best-fit digital-computer program. A fourth-degree polynomial was found sufficient as higher degrees showed no closer agreement to experimental data.

Values of $(E/2)dC/d(a/W)$ were derived from the duplicate 12 × 3-in. specimen compliances. No systematic differences were found between results from the 6, 8, and 10-in. gage lengths of the 24-in. specimen and the length-to-width ratio of 4 was preferable for efficient material usage in test programs. Tabulated values of the dimensionless parameter are provided in Table 4.4 of the section titled "Some Reported Results." Highest accuracy is in the a/W range of 0.25 to 0.40, due to the concentration of data points there.

The experimental calibration compared favorably with the Gross, Srawley, and Brown[9] stress-function calibration. Discrepancies at a/W ratios above 0.4 are probably due to bending in the plane of the specimen which was not taken into account in the boundary-collocation program.

Measurement Precautions

As previously discussed, an ideal compliance measurement involves only that length of specimen which includes all the nonuniform stress field associated with the crack. This procedure allows maximum accuracy and sensitivity since only the displacement due to the crack stress field would be measured over the smallest possible gage length. This was the procedure used in the example previously described.

The geometry of fracture-toughness specimens does not usually allow this simplification, and gage points in most compliance calibrations are positioned to measure the displacement of the actual loading points. Possible measurement errors over those of the ideal case are increased somewhat by longer gage lengths and the added displacement of that portion of the specimen not affected by the crack stress field. More importantly, distortion of loading pins, holes, and clevises are sources of errors. While the deformations associated with the loading region of the specimen are elastic, they are not necessarily linear. For example, in pin-loaded specimens, brinelling and pin flattening involve an increase in supporting area with increasing load. This results in less deformation per pound at the higher loads.

In the case of a specimen loaded in tension by a pin through a hole, the pin clearance at zero load may be lost by elongation and resulting narrowing of the hole as load is applied. The compliance of this region is reduced as the clearance is lost. At the same time, pin clearance is also reduced by pin bending. When the hole interferes, further bending is constrained and measured compliance at high loads is less than at low loads.

The effect of these nonlinearities can be avoided by determining compliance at the same maximum load for all crack lengths investigated. This load cannot exceed that permitted by the longest crack. Therefore, compliances associated with short cracks must be determined from small displacements.

Greater accuracies in compliance calibration can be expected from the larger displacements possible with increased specimen size. This in no way affects the application of the calibration to smaller fracture specimens of similar planar geometry. Material choice for the calibration specimen should be governed by the larger displacements made possible with materials of high yield strength-to-elastic modulus ratio.

The sawed slot necessary to simulate a crack will not provide exactly the same specimen compliance as a closed crack of the same length. Srawley[10] has observed that, although the compliance of a slotted specimen will slightly exceed that of a specimen with an equally long crack, the derivative of the compliance with respect to crack or slot length may be the same. Therefore slot length may be considered the same as crack length. Any uncertainty will be minimized by using large specimens and small slot widths.

Instrumentation

The devices used to measure displacement in compliance calibrations have included dial indicators, Huggenberger extensometers,[4] conventional tension-testing extensometers,[11] and bonded-resistance-strain-gage flexed-beam transducers.[8] The effectiveness of these devices is probably dependent largely on the ingenuity of the investigator as well as the applicability of the measuring device. In choosing a displacement sensor for a specific compliance measurement, some of the basic requirements are sensitivity, accuracy, range, and stability.

The double-cantilever-beam clip gage (Fig. 4.2) recommended by

ASTM[12] for fracture-toughness testing was used effectively in some recent unpublished work by two of the authors. The inherent capabilities of this clip gage were enchanced by the additional instrumentation which comprised the system. Clip-gage excitation was supplied by a commercially available power supply (signal con-

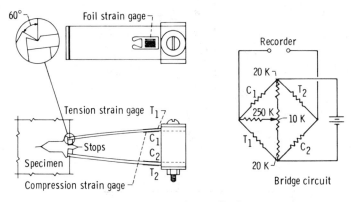

Fig. 4.2—Double-cantilever-beam displacement gage.

ditioner) which proved to be extremely stable. The excitation voltage and gage output were fed through a manual selector switch to a digital voltmeter capable of reading one microvolt. A strain-gage-bridge load cell was energized by another but identical power supply and the excitation voltage and output were also fed through the selector switch to the digital voltmeter. The voltage displayed on this meter could also be recorded on paper tape by a printer activated by a manual switch. The load was applied and held constant by a screw-type machine to insure that the displacement printout could be paired with a known load.

Due to the sensitivity and stability of the digital voltmeter and stability and low noise of the power supplies, the load has been read to ± 1/20th of a pound and displacement to ± 0.000003 in.

Friction Effects

Frictional effects have been examined in the development of fracture-toughness test methods for both bend and compact-tension specimens.[13,14] Loading fixtures (Fig. 4.3) to reduce friction are described in the ASTM Standard E 399-70T: "Tentative Method of Test for Plane-Strain Fracture Toughness of Metallic Materials."[12] Examination of these designs will aid in the preparation of loading devices for calibration of other specimen types. The testing fixtures

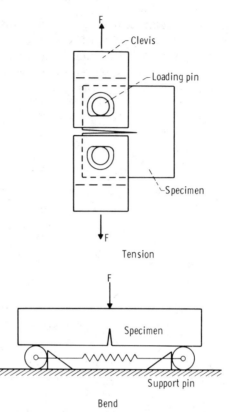

Tension

Bend

Fig. 4.3—Low-friction loading fixtures.

for both types of specimen allow free rolling of pins, both loading pins in the case of the tension specimen and the support pins for the bend specimen. In the bend test, the outward rolling of the pins increased the span with increasing load but this effect is countered by the inward movement of the contact points on the pins with increased bend angle.

Friction effects, as well as the inclusion of extraneous strains, would have no influence on the accuracy of compliance calibrations if they are linear with load. The compliance will be reduced by friction, but the shape of the C vs. a/W curve will not be changed. Friction depresses the curve while the inclusion of extraneous strains elevates it. The derived calibration will not be affected since it is dependent only on the change in slope of the curve.

It is unlikely that the effects of friction are linear with load. For example, consider a loading pin attempting to change its position in a clevis hole as dictated by the change in shape of the calibration specimen under rising load. Initial movement is by rolling, essentially frictionless, but limited by the clearance. When rolling is no longer possible, movement is associated with maximum friction. Therefore the precautions employed in the fracture toughness test fixtures should be observed in design of clevises for compliance measurements.

Data Reduction

The compliances computed from the displacement and load measurements are plotted against the corresponding a/W values. A curve is faired through the points so that $dc/d(a/W)$ can be obtained by graphical differentiation at selected values of a/W. To minimize the error inherent in differentiation, the compliance can be plotted on the logarithmic scale of semi-log paper. A plot of $(E/2)dC/d(a/W)$ vs. a/W is the K calibration curve.

Alternatively, the compliance results for the investigated crack lengths can be fitted to a polynomial in a/W by an appropriate least-squares best-fit digital-computer program. Caution must be exercised because the derivative of the resulting polynomial may not increase monotonically with increasing a/W. This difficulty can be minimized by using the log of the compliance and fitting orthogonal polynomials to the data by the method of least squares.[15] However, judgment is still necessary in selecting the degree polynomial which best represents the data. Too high a degree polynomial produces lower residuals, but at the expense of smoothness in the derivative. In general, experience has shown that third- to fifth-degree polynomials produce sufficiently low residuals and monotonically increasing derivatives.

4.7 RELATED DISPLACEMENT MEASUREMENTS

The experimental determination of the fracture toughness, K_{Ic}, of a material is based on an autographic record relating the load to the displacement across the mouth of the notch at the edge of the specimen. This location is chosen for maximum sensitivity to crack growth. The displacement measured in a K_{Ic} test is not to be confused with displacement measured in a compliance calibration. The gage points for determining a compliance calibration must be positioned such that the displacement measured will be the same as if it were measured at the load axis. In addition, the gage points must encompass the complete change in displacement due to the change in crack length.

4.8 SOME REPORTED RESULTS

Experimental results obtained by compliance calibration are compared with results obtained by numerical methods of analysis in Tables 4.1 through 4.7 for the several different types of plate specimens shown in Figures 4.4(a) through 4.4(g). The sources of these results are Ref. 4 through 9, 11, and 16 through 25 as indicated in the tables.

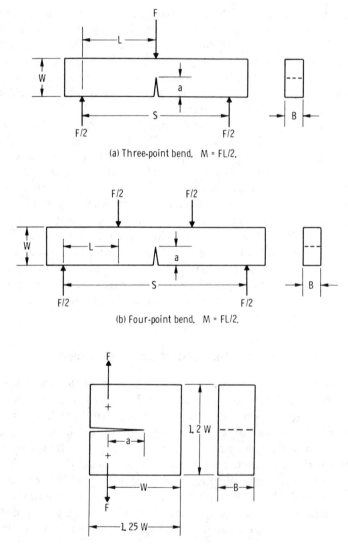

(a) Three-point bend. M = FL/2.

(b) Four-point bend. M = FL/2.

(c) Compact tension.

Fig. 4.4—Compliance calibration specimens.

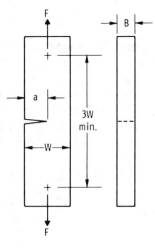

(d) Single-edge-notch tension.

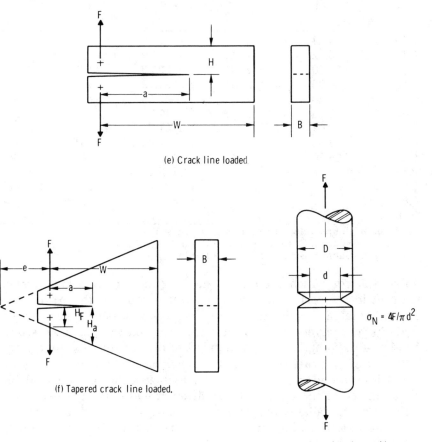

(e) Crack line loaded

(f) Tapered crack line loaded.

$\sigma_N = 4F/\pi d^2$

(g) Notched round bar.

Fig. 4.4—*Continued*

Table 4.1 for three-point bend specimens shows that the ratio of support span to specimen depth has a small but definite effect on the results. The ratio commonly used is 4 (for economy of specimen material) and Table 4.1 shows that there is excellent agreement between unpublished experimental results recently obtained by the present authors and recent boundary-collocation results.[19] These results may be compared with those for four-point bend specimens in

Table 4.1—Three-point Bend Specimen

| | Dimensionless Parameter | | | | |
| | Experimental Results | | | Analytical Results | |
Relative Crack Length, $\dfrac{a}{W}$	*Irwin et al.,[16] $\dfrac{E\mathcal{G}B^2 W}{(M/W)^2}$	†Kies et al.,[17] $\dfrac{E\mathcal{G}B^2 W}{(M/W)^2}$	§ Present Authors, $\dfrac{E\mathcal{G}B^2 W}{(M/W)^2}$	*Gross, Srawley,[18] $\dfrac{K^2 B^2 W}{(M/W)^2}$	§ Gross,[19] $\dfrac{K^2 B^2 W}{(M/W)^2}$
0.10	9.44	10.88	12.79	11.70	. . .
0.15	13.92	17.28	17.99	17.30	. . .
0.20	19.20	24.48	23.49	23.47	21.78
0.25	26.08	33.12	29.79	30.78	. . .
0.30	35.20	41.92	37.58	39.84	36.89
0.35	48.16	57.28	47.76	51.54	. . .
0.40	63.04	. . .	61.75	67.18	62.50
0.45	83.36	. . .	81.60	88.78	. . .
0.50	108.80	. . .	110.8	119.74	112.34
0.55	. . .	. . .	155.2	. . .	. . .
0.60	. . .	. . .	224.9	. . .	225.63
0.65	. . .	. . .	339.1	. . .	. . .
0.70	. . .	. . .	534.5	. . .	545.82

* Ratio of Support Span to Specimen Depth $S/W = 8$.
† Ratio of Support Span to Specimen Depth $S/W = 10$.
§ Ratio of Support Span to Specimen Depth $S/W = 4$.

Table 4.2, for which there is no shearing force or bending-moment gradient in the part of the specimen which contains the crack. The experimental and analytical results for four-point bend specimens are again in good agreement. It should be recalled that, as discussed in Section 4.4, compliance measurement results are directly interpreted in terms of $\mathcal{G}$, whereas results of numerical analysis are usually directly interpreted in terms of K. This is an awkward but unavoidable consequence of the nature of the different methods. The results are directly comparable only if either the value of Poisson's ratio is zero or if the state of deformation is plane stress or plane strain throughout the body, but the extent of the ratio $E\mathcal{G}/K^2$ is only from $(1 - \nu^2)$ to 1.

Table 4.3 shows results for the compact specimen with a ratio of

Table 4.2—Four-point Bend Specimen

Relative Crack Length, $\dfrac{a}{W}$	Dimensionless Parameter		
	Experimental Results	Analytical Results	
	(Lubahn),[4] $\dfrac{E \mathcal{G} B^2 W}{(M/W)^2}$	(Gross, Srawley),[20] $\dfrac{K^2 B^2 W}{(M/W)^2}$	(Bueckner),[21] $\dfrac{K^2 B^2 W}{(M/W)^2}$
0.10	11.8	12.4	12.4
0.15	17.4	18.5	18.3
0.20	24.2	25.3	24.9
0.25	32.15	33.2	33.0
0.30	41.9	42.8	42.9
0.35	53.9	55.2	55.6
0.40	68.6	71.4	72.1
0.45	88.9	92.7	94.5
0.50	118.0	123.0	125.5

Table 4.3—Compact-tension Specimen

Relative Crack Length, $\dfrac{a}{W}$	Dimensionless Parameter	
	Experimental Results (Present Authors), $\dfrac{E \mathcal{G} B^2 W}{F^2}$	Analytical Results (Roberts),[22] $\dfrac{K^2 B^2 W}{F^2}$
0.30	33.0	34.2
0.35	42.6	42.8
0.40	54.8	53.7
0.45	71.6	69.6
0.50	95.9	92.2
0.55	132.7	126.8
0.60	190.4	185.5
0.65	284.5	283.6
0.70	445.3	464.8

H/W of 0.6, Fig. 4.4(c), which is one of the two standard specimens of ASTM Standard E 399-70 T.[12] Again, the results are in very good agreement. Table 4.4 compares results for single-edge notch specimens under remote axial tension. In this case, the experimental results of Ref. 11 were obtained with specimens which were too small for high accuracy. Furthermore, the effect of loading pin friction was neglected in both Ref. 8 and Ref. 11, which might account for some of the discrepancies with the numerical results.[9] It should also be noted that very accurate results can hardly be expected for very short relative crack lengths since there is then a very low rate of change of specimen stiffness with crack length.

Table 4.4—Single-edge-notch Tension Specimen

Relative Crack Length, $\dfrac{a}{W}$	Dimensionless Parameter		
	Experimental Results		Analytical Results
	(Srawley et al.),[8] $\dfrac{E\mathcal{G}B^2W}{F^2}$	(Sullivan),[11] $\dfrac{E\mathcal{G}B^2W}{F^2}$	(Gross et al.),[9] $\dfrac{K^2B^2W}{F^2}$
0.05	0.314	0.35	0.204
0.10	0.556	0.65	0.445
0.15	0.816	1.00	0.758
0.20	1.180	1.40	1.180
0.25	1.735	1.97	1.768
0.30	2.571	2.80	2.603
0.35	3.775	4.20	3.813
0.40	5.436	6.18	5.596
0.45	7.641	8.90	8.276
0.50	10.477	12.50	12.399

Table 4.5—Crackline-loaded Specimen

Relative Crack Length, $\dfrac{a}{H}$	Dimensionless Parameter	
	Experimental Results (Ripling et al.),[23] $\dfrac{E\mathcal{G}B^2H^3}{F^2a^2}$	Analytical Results (Gross, Srawley),[24] $\dfrac{K^2B^2H^3}{F^2a^2}$
2	21.3	21.6
3	17.7	18.0
4	16.1	16.4
5	15.2	15.4
6	14.7	14.8
7	14.3	14.4
8	13.9	14.1
9	13.7	13.8
10	13.5	13.8

Table 4.5 shows results for crackline-loaded (or "double cantilever") specimens, Fig. 4.4 (e), for which the ratio a/W is much less than the ratio a/H, so that the latter becomes the significant variable (so-called semi-infinite specimens). In this case, the experimental results[23] are in excellent agreement with the results of numerical analysis.[24] This is also true of the selection of results for tapered crackline-loaded specimens, Fig. 4.4(f), given in Table 4.6. For further experimental details, the reader should refer to Ref. 6, and for more extensive results, to Ref. 25.

Table 4.7 compares experimental results[4] for the notched round-bar specimen in tension with the analytical results.[5] In this case, the

Table 4.6—Tapered Crackline-loaded Specimen

Relative Crack Length, $\dfrac{a}{W}$	Relative Arm Depth at Crack Tip, $\dfrac{H_a}{a}$	Relative Arm Depth at Load Line, $\dfrac{H_F}{e}$	Dimensionless Parameter	
			Experimental Results (Mostovoy et al.),[6] $\dfrac{E \mathcal{G} B^2 W}{F^2}$	Analytical Results (Srawley, Gross),[25] $\dfrac{K^2 B^2 W}{F^2}$
0.2	0.9615	0.53	119	123
0.3	0.8163	0.53	119	117
0.4	0.7353	0.45	119	117
0.5	0.6667	0.45	119	117

Table 4.7—Notched Round-bar Specimen

Relative Notch Depth, $\dfrac{d}{D}$	Dimensionless Parameter	
	Experimental Results (Lubahn),[4] $\dfrac{E \mathcal{G}}{\sigma_N^2 \pi D}$	Analytical Results (Bueckner),[5] $\dfrac{K^2}{\sigma_N^2 \pi D}$
0.500	0.0529	0.0576
0.600	0.0548	0.0650
0.707	0.0524	0.0671
0.800	0.0471	0.0630
0.900	0.0380	0.0441

analytical results are much more accurate because the high stiffness of this type of specimen makes it very difficult to conduct an accurate compliance calibration.

Recently, the compliance-calibration technique has been applied to a pressurized thick-wall cylinder with a radial crack of uniform depth.[26] Because the crack is of uniform depth (that is, the crack front is straight) the values of $\mathcal{G}$ and K are independent of position along the crack front, so that it is possible to interpret the compliance measurements in terms of the Irwin-Kies relation.[1] For this purpose, the notion of compliance is generalized in a straightforward manner in terms of the derivative of volume with respect to pressure (instead of displacement with respect to force). There are apparently no analytical results available for comparison in this case, and for further details the reader should refer to Ref. 26.

4.9 SUMMARY

The experimental method of compliance calibration can be applied to certain quasi-two-dimensional configurations, such as plate or bar specimens, to obtain a relation for the crack-extension force (or fixed-grip strain-energy release rate with crack extension) $\mathcal{G}$ in terms of load, specimen dimensions, and elastic constants. The method is generally less precise than refined methods of numerical analysis, but it has the advantage that the boundary conditions of load application can be (and should be) practically the same as those which will be used in applications of the specimen to the testing of materials. Numerical methods of analysis suffer from the necessity of simplification of boundary conditions in order that the problem shall be tractable. It is always desirable to obtain and compare results by both the compliance method and a numerical method because their advantages are complementary. The measurement precision needed to obtain good results by the compliance method is no less than that needed for Young's modulus because $\mathcal{G}$ is obtained from the direct measurements by differentiation with respect to crack length, which produces a relative error in $\mathcal{G}$ of an order greater than that in the primary data.

The theoretical basis is not discussed in any detail in this chapter. The original reference is Ref. 1, and more recent and sophisticated accounts are given in Refs. 27 and 28.

REFERENCES

1. Irwin, G. R., and Kies, J. A., "Critical Energy Rate Analysis of Fracture Strength," *Weld. Jnl.* (Res. Suppl.), **33**, 1935-1985 (1954).
2. Irwin, G. R., "Analysis of Stresses and Strains Near the End of a Crack Traversing a Plate," *Jnl. of Appl. Mech.*, **24** (3), 361-364 (1957). [Also, see Discussion in *Jnl. of Appl. Mech.*, 299-301 (June 1958).]
3. Irwin, G. R., "Fracture," *Encyclopedia of Physics* (editor, E. Flugge), **6**, 551-590, Springer, Berlin (1958).
4. Lubahn, J. D., "Experimental Determination of Energy Release Rate for Notch Bending and Notch Tension," *Proc. ASTM*, **59**, 885-913 (1959).
5. Bueckner, H. F., Discussion, "Stress Analysis of Cracks" (Paris, P. C., and Sih, G. C.), *Fracture Toughness Testing and Its Applications*, ASTM STP, **381**, 82 (1965).
6. Mostovoy, S., Crosley, P. G., and Ripling, E. J., "Use of Crackline Loaded Specimens for Measuring Plane Strain Fracture Toughness," Matls. Res. Lab. (1966).
7. Paris, P. C., and Sih, G. C., "Stress Analysis of Cracks," *Fracture Toughness Testing and Its Applications*, ASTM STP 381 (1965).
8. Srawley, J. E., Jones, M. H., and Gross, B., *Experimental Determination of the Dependence of Crack Extension Force on Crack Length for a Single-Edge-Notch Tension Specimen*, NASA TN D-2396 (1964).
9. Gross, B., Srawley, J. E., and Brown, W. F., Jr., *Stress Intensity Factors for a Single-Edge-Notch Tension Specimen by Boundary Collocation of a Stress Function*, NASA TN D-2395 (1964).

10. Srawley, J. E., "Plane Strain Fracture Toughness," *Fracture* (editor, H. Liebowitz), 4, 45-69, Academic Press (1969).
11. Sullivan, A. M., "New Specimen Design for Plane Strain Fracture Toughness Tests," *Matls. Res. and Stds.*, 4 (1), 20-24 (1964).
12. ASTM Standards E 399-70T, "Tentative Method of Test for Plane-Strain Fracture Toughness of Metallic Materials," *Book of ASTM Standards*, Part 31, 911-927 (1970).
13. Brown, W. F., Jr., and Srawley, J. E., "Fracture Toughness Testing Methods," *Fracture Toughness Testing and Its Applications*, ASTM STP 381 (1965).
14. Bubsey, R. T., Jones, M. H., and Brown, W. F., Jr., *Clevis Design for Compact Tension Specimens Used in Plane-Strain Fracture Toughness Testing*, NASA TM X-1796 (1969).
15. Mendelson, A., Roberts, E. Jr., and Manson, S. S. Optimization of Time-Temperature Parameters for Creep and Stress Rupture, with Application to Data from German Cooperative Long-Time Creep Program. NASA TN D-2975 (1965).
16. Irwin, G. R., Kies, J. A., and Smith, H. L., "Fracture Strengths Relative to Onset and Arrest of Crack Propagation," *Proc. ASTM*, 58, 640-660 (1958).
17. Kies, J. A., Smith, H. L., Romine, H. E., and Bernstein, H., "Fracture Testing of Weldments," *Fracture Toughness Testing and Its Applications*, ASTM STP 381 (1965).
18. Gross, B., and Srawley, J. E., *Stress-Intensity Factors for Three-Point Bend Specimens by Boundary Collocations*, NASA TN D-3092 (1965).
19. Gross, B., "Some Plane Problem Elastostatic Solutions for Plates Having a V-Notch," PhD thesis, Case Western Reserve University (1970).
20. Gross, B., and Srawley, J. E., *Stress-Intensity Factors for Single-Edge-Notch Specimens in Bending or Combined Bending and Tension by Boundary Collocation of a Stress Function*, NASA TN D-3820 (1967).
21. Bueckner, H. F., "Weight Functions for the Notched Bar," General Electric Co., Rept. 69-LS-45 (1969).
22. Roberts, E., Jr., "Elastic Crack-Edge Displacements for the Compact Tension Specimen," *Matls. Res. and Stds.*, 9 (2), 27 (1969).
23. Ripling, E. J., Mostovoy, S., and Patrick, R. L., "Measuring Fracture Toughness of Adhesive Joints," *Matls. Res. and Stds.*, 4 (3), 129-134 (1964).
24. Gross, B., and Srawley, J. E., *Stress-Intensity Factors by Boundary Collocation for Single-Edge-Notch Specimens Subject to Splitting Forces*, NASA TN D-3295 (1966).
25. Srawley, J. E., and Gross, B., *Stress-Intensity Factors for Crackline-Loaded Edge-Crack Specimens*, NASA TND-3820 (1967).
26. Underwood, J. H., Lasselle, R. R., Scanlon, R. D., and Hussain, M. A., "Compliance K Calibration for a Pressurized Thick-Wall Cylinder with a Radial Crack," Watervliet Arsenal Rept. WVT-7026 (1970).
27. Bueckner, H. F., "The Propagation of Cracks and the Energy of Elastic Deformation," *Jnl. of Appl. Mech.*, Trans. ASME, 80, 1225 (1958).
28. Rice, J. R., and Drucker, D. C., "Energy Changes in Stressed Bodies Due to Void and Crack Growth," *Intl. Jnl. of Fract. Mech.*, 3, 19 (1967).

5

TESTING SYSTEMS AND ASSOCIATED INSTRUMENTATION

by

S. Roy Swanson
MTS Systems Corporation
Minneapolis, Minn.

5.1 INTRODUCTION

As in other fields of mechanical behavior, fracture mechanics can blame part of its belated emergence on test technique. For example, when one normally carries out a fracture test under constant-loading conditions, the flaw extends very quickly due to the continuous increase in stress-intensity factor associated with the enlarging flaw. In fatigue or stress-corrosion cracking, for example, such constant-loading condition makes the period of propagation of the flaw so short that flaw initiation and failure are considered the same point in life.

The purpose of this chapter is to point out to the readers problems peculiar to fracture testing and to recommend practical solutions to these problems.

While the emphasis will be on the use of closed-loop systems used in actual test systems, it is readily acknowledged that open-loop systems can also be used satisfactorily in many cases. However, it should be noted that the majority of testing in this field is carried out in closed-loop control. This trend is continuing, due to the limited response time of human operators in closing the control loop.

The four basic components to a fracture-mechanics test system are:

1. The basic closed-loop test arrangement as shown in Fig. 5.1
2. The programming units (input)
3. The readout units (X-Y plotters, etc.)
4. The fail-safe units (limit detectors, etc.)

In the following sections, a brief description of each of these basic components will be given.

5.2 A CLOSED-LOOP SYSTEM

Tension-Compression Testing

A typical closed-loop system for tension-compression testing is shown in Fig. 5.1, operating in three possible control modes of stroke, load, and local compliance. The system will automatically maintain the piston at any given point, even if the power source or the actuator load varies depending on the transducer being used. For example, to most precisely control the force applied to the specimen, a load cell (force transducer) is used. To control specimen local compliance, a clip gage would be attached to the specimen. Thus, given adequate force in the hydraulic actuator, the range of the transducer essentially establishes the range of the test.

The specimen is a part of the control loop and, therefore, its

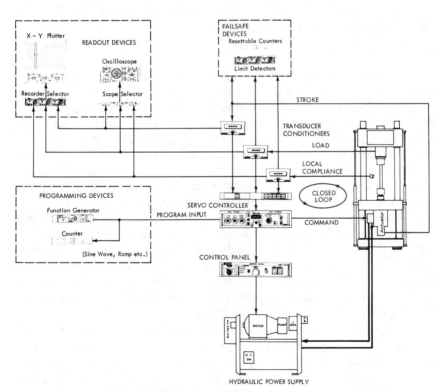

Fig. 5.1—The basic closed-loop test arrangement.

characteristics will directly affect the behavior of the system. Probably the most important property of a specimen, at least as far as its effects on the system are concerned, is its spring rate, because it directly affects the response of the system when operating under dynamic conditions. For example, consider the case of a system which is at zero command level and which is called on by externally generated command to immediately go to a certain load level in tension. In such a case, with a stiffer specimen, the system will approach the command level at a faster rate, increasing the possibility of overshooting the mark. A basic understanding of how the specimen affects system response and stability will be helpful in understanding the reasons and methods for adjusting the servo-controller's gain and stability controls.

In a closed-loop system, the load frame is also part of the loop and, as such, must remain a passive element. If, for example, the loading of the specimen created resonance in the frame, the test would be disrupted with possible damage to the specimen. Generally, the stiffer the frame, the better. Relative motion (deflection) between the load cell and the fixed portion of the actuator (usually the cylinder) should be a minimum.

Pressurization Testing

Another common use of a closed-loop testing system is the pressurization for proof and burst testing of pressure vessels. These systems may be used to control either an internal or confining pressure, and may be programmed to apply static, slowly varying, or high-speed cyclic pressures. The primary variables—maximum static pressures, maximum dynamic pressures, volume expansion of specimen, cyclic rate, and pressurizing medium—cover wide ranges with many possible variations and will, in effect, determine the selection of mechanical components. The advantage of using closed-loop control in a pressurization system is that the closed-loop control automatically compensates for leakage, for changes in the volume of the test specimen, and for changes in the pressure of the power source. Pressurization systems are generally classified into four areas determined by approach: gas direct, hydraulic-fluid direct, and intensifier or booster system. Each of these approaches is described in the following paragraphs.

The gas-direct approach is the simplest and most economical of the three types of pressurization systems. A three-way servovalve under closed-loop control ports gas either from the bottle supply to the specimen or from the specimen to the outlet (see Fig. 5.2). With this system, the flow ranges and volumes can be quite high and the response is good, except at very low specimen pressures or near the

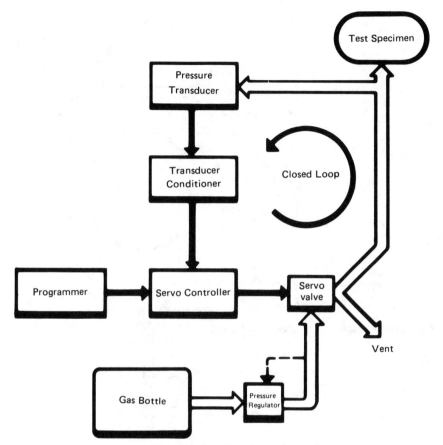

Fig. 5.2—Gas-direct system.

supply pressure. As in all pressurization systems, response depends on the location of the pressure cell with respect to the controlling servovalve and the reaction of the specimen to pressure.

The system shown in Fig. 5.3 uses hydraulic fluid instead of gas. Because fluid is virtually noncompressible, the system response is faster than with gas, the system is safer, and the servovalve has a longer life. Pressure is limited to 6,000 psi by the servovalve, but flow rates of 60 gpm (at 3,000 psi) are routine. Cyclic frequencies of 10 to 20 Hz are common, and frequencies in excess of 100 Hz are possible under optimum conditions. The flow volume is limited by the supply reservoir, and response is excellent at any pressure level except when approaching supply pressure.

The booster system illustrated in Fig. 5.4 is normally used when pressures above 3,500 to 4,000 psi are required or when a conversion from one pressurizing medium to another is required. The magnifica-

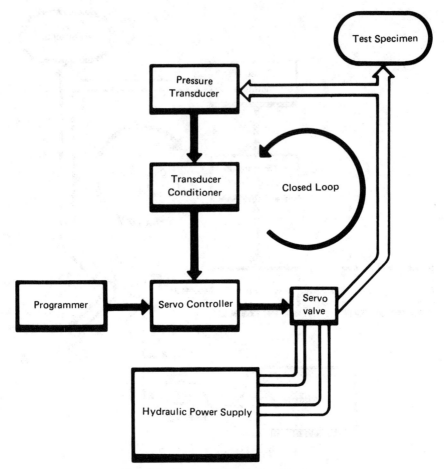

Fig. 5.3—Hydraulic-fluid-direct system.

tion ratio of the intensifier is the ratio of the piston area to the rod
area. Ratios up to 25:1 are commonly used providing pressure outputs from the intensifier of from 10,000 to 50,000 psi. (Pressures as
high as 100,000 psi are possible.) Flow rates at the specimen of up
to 60 gpm are normally used, with flow volumes (limited to the displacement of the high-pressure piston) ranging from 1 to 1,000 cu in.

5.3 PROGRAMMING

For fracture-mechanics testing, there are two program modes:

1. Function generator for fatigue precycling
2. Ramp function for static pull test

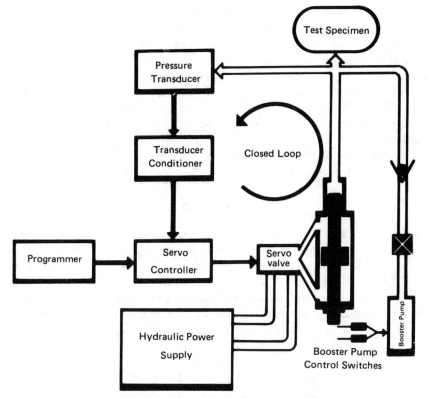

Fig. 5.4—Intensifier or booster system.

Fatigue precycling is carried out in fracture-mechanics studies for the sole purpose of preparing notched specimens in an unequivocal manner for use in the static test to determine the critical value of stress-intensity factor, K_{Ic}. It has been shown that every method of notching specimens, other than in fatigue, produces its own undesirable effects on the value of K_{Ic}.[1] In fatigue precycling, the following various drives are used:

1. Subresonant electrohydraulic
2. Resonant electrohydraulic
3. Mechanical forced vibration
4. Mechanical resonance

Although fluctuating tension loading is usually used for precycling (item 6.5.3, Ref. 1), cycling with reversed loading can be used. For certain tests, such as the circumferentially notched bar, it is desirable to use rotating-beam loading for developing the fatigue crack, since this ensures a uniform crack front. In addition, an initial compressive half-cycle preloading will accelerate the flaw growth, since it will leave a residual-tensile-stress condition at the root of the notch.

Since precycling is secondary to the actual static fracture-toughness test, the accuracy requirements for the dynamic loading are not stringent—usually of the order of 5-percent maximum error (item 6.5.1, Ref. 1). For this reason, it is not necessary to use extensometers or clip-on gages for monitoring the precycling phase. Load control is usually used, although deflection control has the desirable feature for many specimen configurations of a natural drop-off in load amplitude (and, hence, cracking rate) with crack length.

5.4 READOUT

To determine the K_{Ic} value for the specimen, it is usually necessary to produce an autographic record of load vs. displacement of the notch, i.e., the increase in notch width, as the specimen is loaded to fracture in a rising-load test. The load is measured by the load cell which, with the grips, is part of the loading linkage. The notch-opening displacement is measured by a displacement gage whose sensing element is a bonded-wire strain gage. Usually, a single dis-

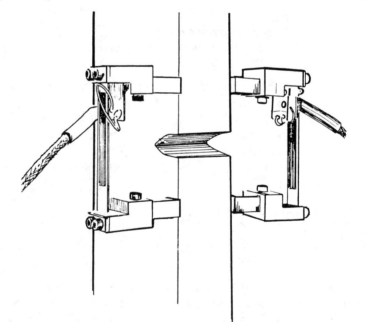

Fig. 5.5—SR-4 electrical-resistance-gage units mounted on SEN specimen.

placement gage, placed at the center of the specimen face is sufficient. However, to determine whether a specimen is being loaded symmetrically and the crack is opening uniformly, additional gages are placed on the face as well as on the specimen sides at the crack front, as shown in Fig. 5.5.

5.5 BASIC DIFFERENCES IN TEST SYSTEMS NEEDED

Introduction

Some years ago, in the study of the static upper and lower yield points for unnotched mild steel, it was observed that the lower yield point was subject to a greater degree of scatter than that of the upper yield point. Eventually, it was shown that the added factor for the scatter was the interaction between the specimen and the testing machine, and that the stiffer the testing machine, the lower the value of the yield point.[2]

We are accustomed to the *static* definition of stiffness as the ratio of applied load to the resulting deflection. In the context of closed-loop-machine characterization, however, *stiffness* is used differently. A stiff testing machine is one that can impose a given external *displacement* regardless of the loads involved. Looking at stiffness in this way, if we command the machine to impose a given rate of displacement, $\dot{\Delta}$, the *loads* that are encountered in moving the grips apart or together at this rate can vary over the complete range of the machine, without causing significant perturbations to $\dot{\Delta}$. It is, then, a relentless or irresistible action of the machine grips or crosshead. This is the ideally stiff or hard machine. A good analogy for this type of action is nut cracking. When we crack a nut, we strive to impose just enough displacement to crack the shell, and the load is removed as quickly as possible to prevent injury to the contents.

There is, however, a contrasting condition, the ideally "soft" machine which imposes a rate of loading, $\dot{P}$, which is applied at all times regardless of the deformations involved. Thus, two conditions are obtained: the ideally stiff machine and the ideally soft machine. All test equipment lie between these two extremes.

As long as a specimen behaves in a completely elastic manner throughout, its load vs. deformation relation is linear, and the behavior of the materials is in phase with the forces brought to bear upon it. For this reason, quantities associated with elasticity, including its limits (such as a sharply defined K_{Ic}, the *onset* of yielding in a flawed structure), will be a function only of the loading program (e.g., the rate of load application, etc.), but not really affected by machine stiffness.

Hard Systems and Soft Systems

So far, the term "stiffness" has been used, but it is recalled that in describing the springy nature of a soft machine, there is a parallel relationship with the response of the machine. *Response* refers to the ability of a testing system to remain faithful to the loading program commanded by the operator.

When working in the purely elastic region for specimens, it is easy to confuse rate of load application with response. The two phenomena are quite different, however, and independent of each other in the yielding region. In fact, high-rate loading tends to minimize the differences in stiffness of systems. They are, of course, very susceptible to frequency response problems in both readout and control.

So far, only machines with open-loop or manual feed-back loading situations have been considered. The physical response of the servo-controlled machine is limited to its appropriate load–amplitude–frequency performance envelope (Fig. 5.6).

Note that a servosystem can be set up with any parameter—load, strain, deflection, velocity, acceleration, etc. In this way, the machine ideally can be made extremely soft (servoed on load) or ex-

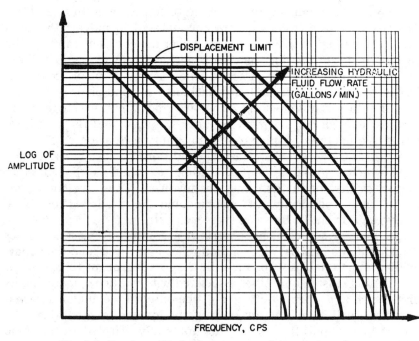

Fig. 5.6—Load amplitude–frequency performance envelope.

tremely stiff (servoed on displacement). In recent cases, the response of closed-loop machines has been extended past 100–200 cps in many applications. Converting back to cycle wavelengths, this refers to about 5–10 msec of response.

At this point, a new quantity will be defined, α;

$$\alpha = \frac{\text{time rate of change of machine parameter}}{\text{time rate of change of specimen parameter}} = \frac{\text{machine response}}{\text{yield velocity}}$$

where we are discussing the same parameter in both cases.

If α is less than unity, the machine response is insufficient to follow the changes which are occurring in the specimen. Experiments using very high-response readout from low-response equipment often show P-Δ relations quite different from that commanded.

Clearly, a machine must have a value of α greater than unity to be in complete control; that is, to remain faithful to the command signal. Due to the near-sonic speeds involved in some tests, α values less than unity will occur during segments of fast crack extension in brittle materials regardless of choice of loading equipment.

Consider the region of post K_{Ic} behavior in fracture mechanics. This is a region of specimen-machine instability. If a test is carried out with servo control on load or load control (we recall that this is equivalent to a soft machine), the load-compliance or load-crack length relation will be obtained (shown in Fig. 5.7). Some point along this curve will yield a value of K_c, defined as the value of stress-

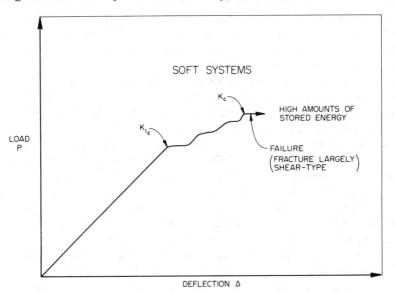

Fig. 5.7—Soft-system characteristics.

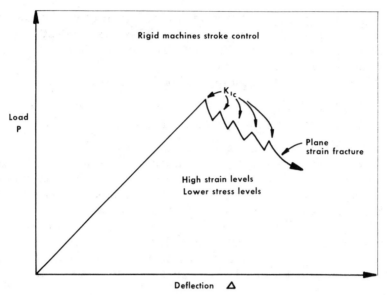

Fig. 5.8—Stiff-system characteristics.

intensity factor associated with "unstable" or fast fracture. This is an important parameter in *thin*-sheet fracture-toughness testing.

Consider now the same test carried out with servo control based on crosshead travel or displacement of the grips. This may be called stroke control since actuators are employed which can readily be switched from load control to stroke control; this corresponds to a hard or stiff machine. In this case, the load-compliance curve for certain specimens (crackline loaded, compact K_{Ic}) will have the following shape, as shown in Fig. 5.8.

These peaks occur after crack arrests and are, in reality, a form of K_{Ic} measurement associated with the particular elasticity effects generated by the arrest of running cracks (slightly higher than "normal" K_{Ic} measurements). Figure 5.9, taken from Ref. 3, shows this difference between hard and soft systems for arrest characterizations.

Now the question arises: What will be the value of K_c in the thin-sheet fracture-toughness testing? Under certain conditions, e.g., specimens with pronounced slow crack growth and appropriate specimen design, values of K_c are largely a function of α. An ideally stiff machine with zero response to load would not even obtain K_c in some cases with wedge-loaded specimens! (Fig. 5.8.)

In many open-loop machines, there is a single-valued relationship between loading rate and stiffness during dynamic testing. An increase in cyclic frequency, for instance, can cause an increase in system stiffness, since the machine inertia loads are acting to reinforce

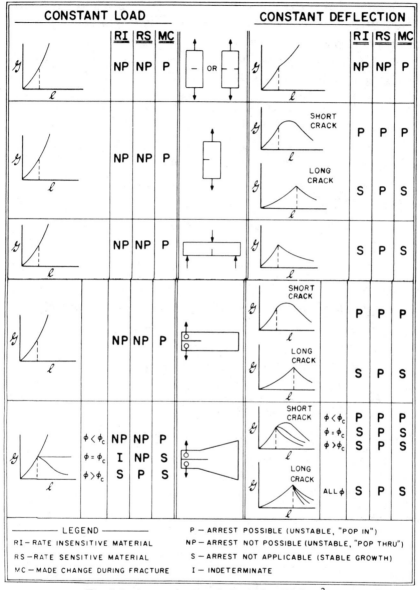

Fig. 5.9—Arrest characteristics (After Bluhm.[3])

the programmed loading. At resonance, we obtain a "soft" condition since there can be an increased deflection for the same loading with decreased specimen stiffness.

The relation between crack-extension force G and stress-intensity factor K is not affected by these considerations, as shown below.

Assume that a specimen containing a crack is loaded to P_o. At this *point*, assume that the flaw advances δa in length:

(a) *Stiff System*

The load will immediately decrease to $P_o - \delta P$, as the displacement (compliance) increases to $e_o + \delta e$.

$$\mathcal{G} = \lim_{\delta A \to 0} \left\{ \frac{\delta W' - \delta U}{\delta A} \right\}$$

$$= \lim_{\delta A \to 0} \left\{ \frac{\frac{1}{2}(P_o - \delta P)(e_o + \delta e) - \frac{1}{2} P_o e_o}{B \delta a} \right\} \tag{5.1}$$

Compliance $C = e/P$

Thus,

$$\delta e = P \delta C + C \delta P$$

Thus,

$$\mathcal{G} = \lim_{\delta A \to 0} \left\{ \frac{P_o^2}{2} \frac{dC}{B \delta a} \right\} = \frac{P_o^2}{2} \frac{1}{B} \frac{\partial C}{\partial a} \tag{5.2}$$

where:

A = area, W' = work done, U = strain energy, B = width.

(b) *Soft System*

As before,

$$\mathcal{G} = \lim_{\delta A \to 0} \frac{\delta W - \delta U}{\delta A}$$

but, now, there is no change in load level P_o

$$\mathcal{G} = \lim_{\delta A \to 0} \left\{ \frac{\frac{1}{2} P_o (e + \delta e) - \frac{1}{2} P_o e_o}{B \delta a} \right\}$$

$$\mathcal{G} = \lim_{\delta A \to 0} \frac{\frac{1}{2} P_o \delta e}{B \delta a}$$

As before,

$$\delta e = C \delta P + P \delta e$$

But

$$\delta P = 0$$

Thus,

$$\delta e = P \delta C$$

and

$$\mathcal{G} = \lim_{\delta A \to 0} \frac{P_o^2}{2B} \frac{\partial C}{\partial a} \tag{5.3}$$

as found in eq. (5.2).

At the present, a good basis for K_{Ic} testing has been established (e.g., Ref. 1), and the activity of various research centers is shifting to the more complex mixed-mode thickness-dependent fracture associated with determinations of K_c. Three factors affect K_c:

1. Specimen geometry
2. Machine response (hard or soft system)
3. Loading speed (interaction with material relaxation)

Specimen geometry involves other factors besides thickness. The eccentricity of the load path between grips has a great effect on post K_{Ic} behavior (Fig. 5.10). The essential point is that, if we have remote (concentric, axially aligned, etc.) loading, then the stress-intensity factor in a centrally notched plate increases. However, with wedge-loaded specimens

$$K = \frac{\sqrt{2}P}{\sqrt{\pi a}} \tag{5.4}$$

Thus, as crack length increases, K actually decreases and crack arrest can occur.

It is possible with stiff systems for a specimen to fail by plane-strain fracture (K_{Ic}), which also is prevalent with the same specimens in soft systems. For K_c tests, then, α should be evaluated for assurance that the response is adequate for the test. One way to achieve this is to use a two-channel oscillograph of very high response (such as with a storage oscilloscope or an optical-galvanometer system), with one channel recording the load and the other recording the specimen yield. The latter quantity can best be obtained from the local cracking or clip-gage signal. If all elements of this system have

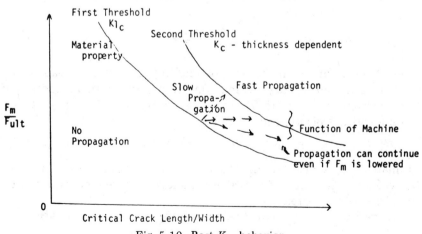

Fig. 5.10—Post K_{Ic} behavior.

responses of at least 1,000 cps, we can determine if our test for K_c is really valid.

What stiffness or response is appropriate in service conditions? This is not so important perhaps with brittle materials, as it is with ductile materials. However, as fracture mechanics is being extended down to tougher materials at present, test system response and the appropriate service load-shedding conditions are matters of considerable significance.

Testing Problems (after Irwin[4])

In the determination of K_c by large-sheet testing for application to aircraft-fuselage-sheet materials, the area of primary interest seems to be in the range of $(2r_{ys}/B)$ values, where 100-percent shear fractures are expected. For such fractures, the limiting crack speed should be *depressed* below 1,000 ft/sec, and it is not clear that we can rely on an abrupt velocity jump to provide an unambiguous measurement point. A stable crack-extension speed on the order of 0.1 in./sec may be fast enough so that the load-relaxation rate due to the crack speed balances the load-increase rate expected from the speed of separation of the specimen grips, $V = dl/dt$, and produces a maximum load point. This separation speed can be expressed as:

$$C = \text{compliance}$$

$$V = C\frac{dP}{dt} + P\frac{dC}{da}\dot{a}$$

where P = applied tensile load.

When $dP/dt = 0$ and V is assumed constant, one finds

$$\dot{a} = \frac{1}{t}\frac{d}{da}(\log C)^{-1} \qquad (5.5)$$

where t is the time from start of loading to the maximum load point.

Under the customary testing conditions, only a small fraction of the crack speed in eq. (5.5) would be contributed by dr_{YS}/dt. Assuming width = plastic-enclave radius

$$\dot{a} = n\,dr_{ys}/dt \qquad n = \text{constant} \qquad (5.6)$$

estimates for typical testing situations show n must be in the range 30 to 200. The value of n increases rapidly with reduction of the ratio of net section stress to σ_{YS}.

If we hypothesize that a velocity jump occurs in the relationship of $\dot{a}$ to $\mathcal{G}$ when a certain critical value of stable crack speed has been achieved, the timing of this condition can be placed either before or after the maximum load point by changes of specimen size (or shape). In addition, the various factors which contribute to the stable crack speed are not well understood at the present time. Thus, the basic uncertainty factors, which must be clarified to some degree to provide a rational basis to writing a testing standard, are those concerned with stable motion of the crack under K_c testing conditions.

Other aspects, although less basic, are also in need of investigation. Presumably, we could measure effective crack size in terms of apparent compliance with a suitable displacement gage. This would supplement direct observations of crack size and would eliminate the need for a plasticity adjustment factor. However, some experimental work will be needed to demonstrate how this should be done. Presumably the test sheet can be held sufficiently flat so that the stress analysis does not need correction factors for buckling or warping of the test specimens. However, methods for doing this also need investigation.

Crack-extension Resistance (after D. P. Clausing[5])

A stable tear. If the initial increment of crack extension causes $\mathcal{G}_{resistance}$ to become larger than $\mathcal{G}_{applied}$, the crack will not propagate farther until an additional change is imposed on the test system; thus, the crack is stable.

An unstable tear. If the initial increment of crack extension causes $\mathcal{G}_R$ to become smaller than $\mathcal{G}_A$, the crack will continue to propagate, even if no additional change is imposed on the test system; thus, the crack is unstable.

In a plate with a through-thickness crack, the crack-tip applied load is given by

$$\mathcal{G}_A = \frac{1}{2B} p^2 \frac{dC}{da} \qquad \text{Soft Systems} \qquad (5.7)$$

or

$$\mathcal{G}_A = \frac{1}{2B} \Delta^2 \frac{dc}{da} \qquad \text{Stiff Systems} \qquad (5.8)$$

Equation (5.7) is preferred when the *force* is controlled, and eq. (5.8) is preferred when the *displacement* is controlled. Three definitions are

$$K_A = Y \frac{P}{BW} a^{1/2},$$

$$f_2 = \frac{2}{Y} \frac{dY}{d(a/W)} + \frac{1}{a/W}$$

$$f_3 = \frac{4MY^2\, a/W}{EB\lambda} \tag{5.9}$$

where W is the plate width, and Y a dimensionless geometrical parameter. Differentiating eqs. (5.7) and (5.9) and rearranging, one obtains

$$\frac{da}{dP} = \frac{2\mathcal{G}_A\, W}{P\left\{\dfrac{d\mathcal{G}_A}{d(a/W)} - f_2\, \mathcal{G}_A\right\}} \tag{5.10}$$

and

$$\frac{da}{d\Delta} = \frac{f_1\, MEW\, \mathcal{G}_A}{\dfrac{d\mathcal{G}_A}{d(a/W)} - (f_2 - f_3)\mathcal{G}_A}. \tag{5.11}$$

When the crack starts propagating, $d\mathcal{G}_A/d(a/W)$ becomes finite and, in some situations, the value of the denominator in eqs. (5.9) and (5.10) can decrease to zero. When the denominator of eq. (5.9) becomes zero, the crack continues to propagate, even though dP is zero. For a crack in a load-controlled soft system, this is the definition of instability. The instability criteria are

Criterion	Crack Unstable in:	
$\dfrac{d\mathcal{G}_A}{d(a/W)} - f_2\, \mathcal{G}_A = 0$	Load-controlled system $(dP = 0)$	(5.12)
$\dfrac{d\mathcal{G}_A}{d(a/W)} - (f_2 - f_3) = 0$	Displacement-controlled system	(5.13)

The value of a crack-tip load when the crack becomes unstable, $\mathcal{G}_c$, is not only a function of the plate material and thickness and fracture mode, but also depends on the specimen geometry and size, and on the *compliance of the loading system*. The crack-tip resistance, $\mathcal{G}_R$, on the other hand, is essentially a property of the plate material and thickness and fracture mode, if the crack propagation is time independent. Once $\mathcal{G}_R$ has been experimentally determined as a function of crack-propagation distance for a particular plate material and thickness and fracture mode, the value of $\mathcal{G}_c$ can be calculated for the same material and thickness and fracture mode for any plate configuration for which the elastic stress analysis is known.

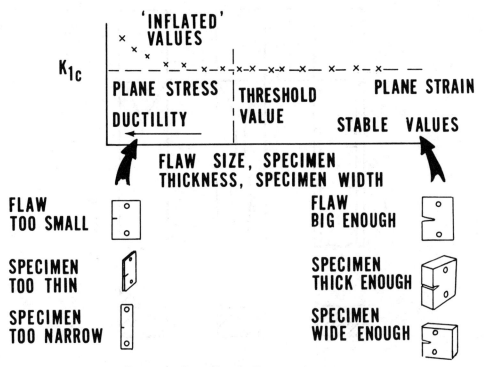

Fig. 5.11—Size effect in K_{Ic} measurements.

The Need for Large-scale Tests

To establish values of K_{Ic}, which is as close to a material property as any quantity in fracture-toughness testing, one must obtain fractures under plane-strain conditions. The amount of material which yields at the crack tip must also be small. To ensure this, specimens must be of sufficient thickness so that a triaxial state of stress can exist at the flaw tip. The need for "plane strain" or thick-plate testing is not a trivial one. K_{Ic} can vary over a range of three to one if the influence of crack length is eliminated. Crack-toughness values vary from a high in thin plates to a low in thick plates. The low value appears to be an asymptote (Fig. 5.11); further reductions in toughness do not occur in thicker plates.

As a result, the view has been adopted by many that the thick-plate asymptote of crack toughness is indeed a material property. The symbols K_{Ic} and G_{Ic} have, thereby, been restricted to denote only this asymptote, commonly called in the literature the "plane-strain values." Note that the term "plane strain" in this context does not mean precisely the same thing as in two-dimensional elasticity; the association is drawn from the thickness of the plate specimens.

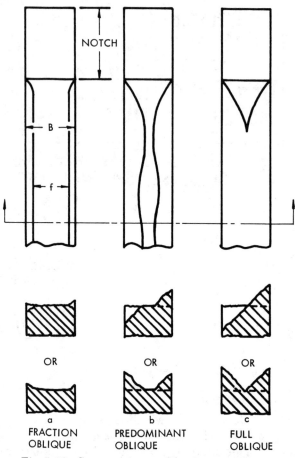

Fig. 5.12—Common types of fracture appearance.

While such a definition might appear to be contrived, it should be borne in mind that "brittle" fracture proceeds at low values of applied stress (Fig. 5.12). In contrast, increasing amounts of yielding can be associated with greater absorption of plastic energy and therefore—in terms of our model—higher levels of the fracture stress. As a result, enforced brittleness of the fracture test gives values of crack toughness which are low, and are interpreted to be conservative.

Without reviewing the details of the requisite arguments, we may note that some tentative specifications for experiment have been developed. A survey of a broad range of laboratory tests indicates that, for valid data, plate thickness must be greater than $2.5 \ (K_{Ic}/\sigma_y)^2$ where σ_y is the uniaxial yield stress. Procedures[6] are outlined for the entire experiment so that tests on identical specimens performed in

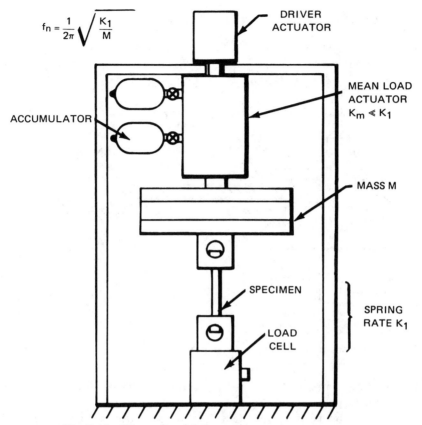

$$f_n = \frac{1}{2\pi}\sqrt{\frac{K_1}{M}}$$

DRIVER ACTUATOR

MEAN LOAD ACTUATOR $K_m \ll K_1$

ACCUMULATOR

MASS M

SPECIMEN

SPRING RATE K_1

LOAD CELL

Fig. 5.13—Schematic of fatigue system using resonance.

different laboratories will yield essentially the same values of crack toughness. These procedures apply to a number of specimen geometries; work in progress is directed toward application to other shapes.

Very large-scale tests using resonance. Since large-specimen tests are necessary, and fatigue precycling should be carried out as quickly as economically possible, we have the requirement for high-dynamic-load capability at "high" frequencies. The basic advantages of the resonant system for certain fatigue-testing applications are much lower operating costs and, possibly, lower initial cost than for a conventional system.

There are two types of servo–hydraulic test machines using resonance which have been developed. The first type is a direct spring-mass design for operation at loads up to about 150,000 lb. It is set up with a specimen as the spring, and some large weights as the mass. As the specimen size increases, the specimen spring rate increases,

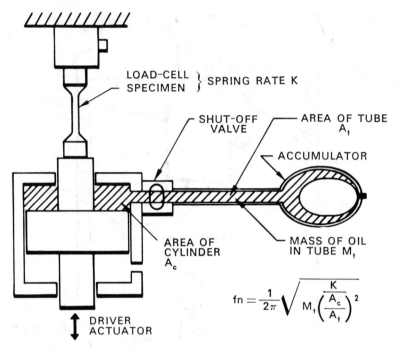

Fig. 5.14—Schematic of fatigue-test system using a hydraulic lever (Russenberger principle).

thus making it necessary to increase the mass size enormously to provide valid test results.

The second type of resonant test machine uses a mass magnification technique where a small mass simulates a large mass. This can be thought of as a mass lever system which has been used in older test machines for some time, except that we are using a hydraulic lever, as shown in Fig. 5.13.

In Fig. 5.14, the principle used for a mass magnification is shown as a hydraulic lever. In this case, the "lever arm" is the ratio between the hydraulic actuator area and the cross-sectional area of a long, thin tube.

Due to the fact that these systems are resonant, they can operate at only one frequency for a given spring-mass combination. Since the spring is the specimen being tested, the mass must be varied to get a range of operating frequencies for these systems.

Fracture-mechanics Tests Involving Environmental Chambers

To establish the effect of temperature on K_{Ic} values, specimens are often tested over the range of temperature of interest. To meet this requirement, low-temperature chambers or furnaces are often built to provide the needed range of temperature. The cooling system used in the cold chamber can be, for example, liquid-nitrogen spray. It is chosen because it has several advantages over a compressed gas system. The initial cost is only a fraction of the cost of a compressed-gas system. The heat-extraction rate can be increased as desired simply by increasing the rate of liquid-nitrogen spray. Finally, the system is light, very portable, and easily adaptable.[7]

A Simplified Fracture-mechanics Test

Many studies in the related area of stress corrosion or subcritical crack growth use simplified test systems, many of which are truly elegant in their simplicity as shown by the example in Fig. 5.15.[8] Placed in a corrosive environment, such a preloaded specimen will give valid answers to the question of onset of cracking in a stressed

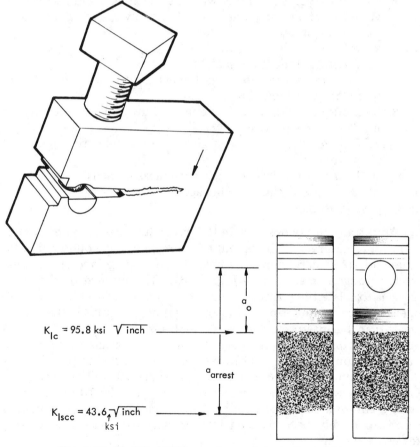

Fig. 5.15—Modified WOL specimen of steel after arrest.

structure. Proving rings have also been used for the same type of test.

One can observe "pop-in" with the specimen immersed in a corrosive fluid, watch the crack grow, and then measure the length of the crack. The deflection of the ring gives the load, and the crack length permits one to measure K.[8]

5.6 MONITORING CRACKING AT THE NOTCH

Introduction

We can summarize the different methods of crack-length monitoring grouped in order of increasing desirability, into the following sections:

1. Methods involving contact with the specimen (often with individual specimen preparation) and possibly affecting the cracking process it detects:
   a. Ink stain
   b. Coatings (thermal, visual, etc.)
   c. Bulk measurements (e.g., high-current resistance measurement by milliohmeter)
2. Methods involving specimen contact but, probably, not affecting the fatigue process:
   a. Voltage drops across a given section (electrical-potential method)[7]
   b. Low-current (bridge) resistance measurement
   c. Compliance gages, strain gages
   d. Ultrasonics

An ultrasonic nondestructive-test procedure has been developed[9] to measure and record the extent of crack growth encountered in fatigue and stress corrosion tests involving the wedge-opening-loading fracture-toughness specimen (Fig. 5.16). The essence of the technique is to relate the position of an ultrasonic transducer on the specimen surface to the tip of the propagating crack such that crack length can be interpreted in terms of transducer location. The required instrumentation includes commercially available ultrasonic flaw-detection equipment and a test fixture designed to permit completely automatic measurement of crack growth. The technique yields a crack-length-measurement sensitivity of ±0.010 in. and provides a continuous record of crack length vs. elapsed time which, in turn, can readily be converted into crack-growth-rate data suitable for use in design.

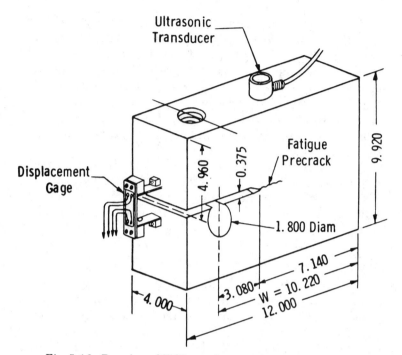

Fig. 5.16—Drawing of WOL specimen with ultrasonic probe.

3. Methods not touching the specimen, but not readily amenable to continuous crack monitoring:
   a. Visual observances. A typical but crude method used in monitoring cracking is to use a large magnifying glass, sometimes with illumination, either constant or strobo-scopic. Some such devices have built-in micrometer scaling, and quite good measurements (to .002 in.) can sometimes be obtained without stopping the test.
   b. X rays
   c. Certain optical (interference) methods

4. Methods not touching the specimen and amenable to continuous and automatic crack-length recording:
   a. Optical servomechanisms
   b. Surface-crack detection by microwave methods
      This surface-crack detection system that can be used to examine metallic surfaces with a noncontacting probe has application to the continuous observation of the growth of flaws during fatigue or fracture tests. A microwave metal-surface-flaw detector is used to irradiate a flawed metal

surface with electromagnetic microwave energy. This ir-radiation results in re-emission of electromagnetic energy from the surface in a pattern of eigenmodes different from those of the original irradiating signal. The generation of a spatial periodic signal with a polarization modulator in-duces a characteristic received pattern which is correlated either with an inserted reference pattern or autocorrelated with itself to detect a change in surface properties of the test specimen.

It has been demonstrated that this microwave tech-nique can detect flaws and scratches as small as 100 μ in. The system is noncontacting, nondestructive, and poten-tially small enough to be hand carried over a surface under tests.

Eddy-current Probes (with Lift-off Compensation)

The eddy-current test, which appears quite promising, can be de-scribed physically as follows. When a coil carrying a high-frequency current is placed near an electrical conductor, such as the aluminum-plate specimen, eddy currents are induced in the specimen. By Lenz's law, there is also a magnetic field associated with this induced current. Flaws can cause a local change in resistance which alters the current pattern and, hence, alters the associated magnetic field. De-tection and monitoring of this magnetic field will, therefore, detect a crack.

In a testing program, a point probe will interrogate the small vol-ume of material beneath it. By placing the probe at the edge of, say, the center notch in the specimen, and setting up a servoactuator to move the probe horizontally upon indication of a crack, the detector can be "locked on" to the crack tip.

The most critical factor is the distance from probe to specimen. A likely solution to this problem, where specimen vibration is pres-ent, is to recess the probe into a hard-rubber sheath, with the dis-tance from the tip of the sheath to the probe face kept constant (at, say, 0.005 in.). It is often difficult to have sufficient contrast dis-crimination to keep visual track of the tip of a fatigue crack on the surface of uniform reflectivity. This problem is handled very well by eddy-current devices, which see the crack as a large perturbation con-trasting strongly with the background of uncracked sheet. Also, they have reasonable power to penetrate the top surface layers to some degree, to obtain an accurate and stable signal.

The eddy-current method is essentially a combination of a mag-netic method and a local resistance-measuring technique. Its advantages include: automation is simple and no special coupling problems are involved.

This method promises to result in a very precise measure of crack length. It is continuous, and the servosystem can be arranged for automatic crack following, as shown in Fig. 5.17.

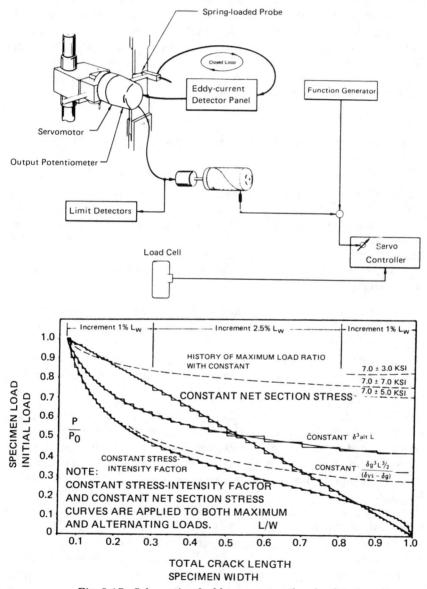

TOTAL CRACK LENGTH
SPECIMEN WIDTH

Fig. 5.17—Schematic of eddy-current probe circuit.

Another application of closed-loop eddy-current probes is in the actual evaluation of the critical plane-strain fracture-toughness parameter K_{Ic}. By increasing the sensitivity of the balance in the servo-system, and using frequencies appropriate for deep penetration, one can detect the "pop-in" event more accurately, presumably, than with any other system used so far, at least for thin materials. When one considers the comparatively low sensitivity with a compliance system at low percentages of crack length, where very little change in compliance occurs in the initial 5 percent of crack length/width, the preparation associated with accurate compliance readings, and their greater vulnerability to temperature, the servo-controlled eddy-current probe has definite advantages.

5.7 AUTOMATION OF VARIOUS COMPONENTS OF THE TEST

Fully Automated Computer-controlled Fracture Mechanics Tests

Lately, closed-loop electrohydraulic systems have been controlled by a digital computer for carrying out fracture-mechanics tests. Since the future will undoubtedly see more use of computers in the test laboratory, it is instructive to describe in detail what changes this makes in the test procedures already discussed.

There are two main areas of fracture-mechanics testing where computers are useful:

1. The programming of several test systems concurrently in a large laboratory, for instance, to generate K_{Ic} information in quantity.
2. The carrying out of tests at high speeds to obtain proper simulation of service effects, where operator response is not sufficient.

The latter point may sound like the argument used earlier for using closed-loop principles. However, the computer can also keep track of instantaneous K values, and can alter much more quickly than any other method can, such quantities as fatigue precycling to suit prearranged limits of compound parameters, e.g., load $\times$ crack length$^{1/2}$.

Using the computer also permits the machine to make decisions regarding secant slopes of K_{Ic} curves, etc. (in an objective manner), which are quite tedious for the person who must analyze fracture-mechanics K_{Ic} records. A typical computer-controlled fracture-mechanics program precracks a specimen by cycling in load control between specified limits until the desired crack length is obtained,

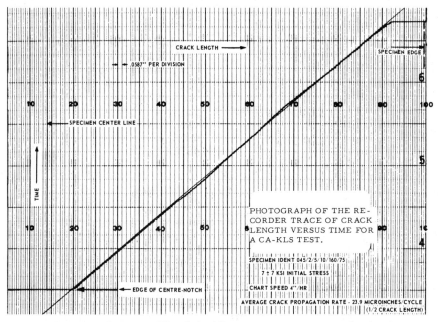

Fig. 5.18—Constant K-load shedding test at constant amplitude.

and then pulls the specimen apart to calculate the K_{Ic} (infinite width) and the maximum load.

Crack Growth Using Continuous Crack-length Signal

Recently, a method was developed[10] to obtain a continuous electronic reading of the crack length by use of a servo-controlled eddy-current probe. Applying appropriate finite-width correction factors, one can construct load–crack length histories to follow "constant K" profiles. That is, the load is dropped off in proportion to the crack length (which is supplied as a continuous signal) to maintain a constant amplitude of stress-intensity factor during crack growth in the specimen. When a constant K curve was used for fatiguing 7079-T6 aluminum alloy,[10] an essentially straight line on the crack length-time plot was obtained (Fig. 5.18).

5.8 SUMMARY AND CONCLUSIONS

In this chapter, we have tried to create an awareness of those factors which can compromise the validity of fracture-mechanics experimental data. Two main trends have been highlighted, namely, the use of closed-loop testing, and the increasing application of automation to fracture-mechanics testing.

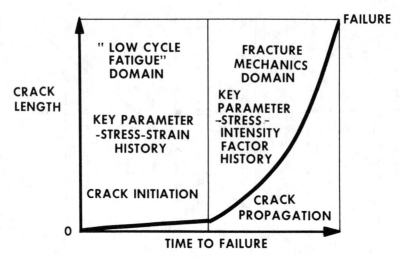

Fig. 5.19—Crack growth.

Many areas have been discussed only briefly due to the limitations set by a consistent treatment in this monograph. The list of references should form a starting point for readers who are getting involved in the experimental aspects for the first time.

Finally, it should be pointed out that in any problem involving the life prediction of a part, the discipline of fracture mechanics is the primary tool for the latter stage in the process (see Fig. 5.19)—the growth of the existing flaw. The initial stage—the incubation of the flaw—is either assumed a trivial problem (the flaw is there to begin with) or comes about in the manner currently best described by a continuing analysis of stress and strain histories, as indicated in Fig. 5.19.

REFERENCES

1. *Tentative Method of Test for Plane-Strain Fracture Toughness of Metallic Materials*, ASTM Tentative Standard E-399-70T (Mar. 1970).
2. Brittain, J. O., "Modern Applications of the Theories of Elasticity and Plasticity to Metals," *Applications Bulletin—The Instron Corporation.*
3. Bluhm, J. I., "Fracture Arrest," *Fracture, an Advanced Treatise* (editor, H. Liebowitz), **5**, 1–63 (1969).
4. Irwin, G. R., "Fracture," *Handbuch der Physik*, **6**, 551–590 (1958).
5. Clausing, D. P., "Crack Stability in Linear Elastic Fracture Mechanics," *Intl. Jnl. of Fract. Mech.*, **5** (3), 211–227 (Sept. 1969).
6. Brown, W. F., and Srawley, J. E., *Plane Strain Crack Toughness Testing of High Strength Metallic Materials*, ASTM STP 410 (Dec. 1967).
7. Spewock, M., "Servo-Controlled, Electro-Hydraulic System for Testing

Fracture Toughness Specimens," Westinghouse Res. Lab., Res. Rept. 69-1P96-BTLFR-R3 (Aug. 5, 1969).

8. Novak, S. R., and Rolfe, S. T., "Modified WOL Specimen for K_{Iscc} Environmental Testing." *Jnl. of Matls.*, 4 (3), 701–729 (1969).

9. Clark, W. G., Jr., and Ceschini, L. J., "An Ultrasonic Crack Growth Monitor," Westinghouse Res. Lab., Scientific Paper 68-7D7-BFPWR-P1 (Apr. 8, 1968).

10. Swanson, S. R., Cicci, F., and Hoppe, W., "Crack Propagation in Clad 7079-T6 Aluminum Alloy Sheet under Constant and Random Amplitude Fatigue Loading," *Fatigue Crack Propagation*, ASTM STP 415, 312–362 (1967).

6
PHOTOELASTICITY TECHNIQUES

by

Albert S. Kobayashi
University of Washington
Dept. of Mechanical Engineering
Seattle, Wash.

6.1 INTRODUCTION

The purpose of this chapter is to demonstrate the usefulness of photoelasticity in solving boundary-value problems in fracture mechanics. Detailed descriptions of photoelasticity, as well as various photoelastic techniques, are to be presented in separate SESA monographs and are, therefore, omitted in this chapter.

Two-dimensional photoelasticity has been used in the past to analyze boundary-value problems in fracture. Dixon studied the influence of finite-width effects on doubly edge-notched and centrally notched plates[1,2] by photoelasticity and the elastic-plastic strain distributions in metal plates using birefringent-coating technique.[3-5] Gerberich also used birefringent-coating technique to investigate the strain fields in cracked aluminum plates.[6] A recent paper by Hubbard reports the use of birefringent coating in studying the plastic-strain distribution in fatigue-crack growth in centrally notched aluminum 7075-T6 plates.[7] Van Elst[8] and Pratt[9] used the birefringent-coating technique with ultra-high-speed photography to record crack propagation and the associated strain field in steels. Recently, Dixon used two-dimensional photoelasticity to study the buckling deformation and associated stress field around centrally notched thin plates.[10] Wells and Post, in a widely quoted paper, studied fracturing process in CR-39 photoelastic plastics by the use of two-dimensional dynamic photoelasticity and photoelastic interferometry.[11] This procedure was further extended in a recent paper by Bradley.[12] Details of the latter work will be discussed in 6.3.

Leven investigated variations in the three-dimensional state of stress at crack fronts of compact tension fracture-toughness specimens[13] and in the vicinity of surface flaws in pressure vessel.[14]

Photoelastic models were constructed from epoxy resin and the frozen-stress technique was used to determine the maximum-shear-stress distribution in slices perpendicular to the crack front. Variations in the stress-intensity factor along the crack front were also computed.

When photoelasticity is used to determine stress-intensity factors in the above problems, the severe stress concentration at the sharp crack tip in the photoelastic model causes plastic yielding at the crack tip and thus the observed birefringence cannot be related directly to the state of stress in such region.* In addition, the extreme contrast of high fringe orders at the crack tip and low fringe orders away from the crack tip in the photoelastic model makes it difficult to select a model material which would be ideally suited for the entire region.† Thus, some modifications in conventional photoelasticity are necessary when problems in fracture mechanics are analyzed.

In the following section, these special techniques in photoelasticity are discussed. Particularly, emphasis is placed on certain problems in fracture mechanics for which photoelasticity offers the best solution technique available to date.

6.2 TWO-DIMENSIONAL STATIC PHOTOELASTICITY

Isothermal Analysis

The direct approach for determining stress-intensity factors is through the use of eq. (2.34). The maximum in-plane shear stress derived by using these equations is

$$\tau_m = \frac{1}{2\sqrt{2\pi r}} \{[K_I \sin \theta + 2K_{II} \cos \theta]^2 + [K_{II} \sin \theta]^2\}^{1/2} \qquad (6.1)$$

where τ_m is determined directly from the isochromatic patterns in two-dimensional photoelasticity. The obvious choice then is to measure τ_m at various locations of (r,θ) within the region surrounding the crack tip and compute an average value of K_I and K_{II} by the use of eq. (6.1).

Smith and Smith,[18] following Irwin's discussion in Ref. 11, sug-

---

*Perhaps this obstacle can be overcome with future advances in photo-plasticity.

†This difficulty can be overcome by using standard techniques of increasing the optical sensitivity, such as fringe multiplication[15,16] and Tardy compensation[17] techniques to measure fractional fringe orders in regions away from the crack tip.

gested that the angle, θ_m, where τ_m is maximum for given r be used for separating K_I and K_{II} in eq. (6.1). By setting $\partial \tau_m / \partial \theta = 0$,

$$\left(\frac{K_{II}}{K_I}\right)^2 - \frac{4}{3}\left(\frac{K_{II}}{K_I}\right) \cot 2\theta_m - \frac{1}{3} = 0 \qquad (6.2)$$

is obtained which enables one to determine the ratio of K_{II}/K_I once θ_m is measured from the isochromatic patterns. In practice, however, this ratio is sensitive to the experimental variation of θ_m as discussed by Bradley[12] but, fortunately, this experimental scatter has lesser effect on K_I which is the dominant mode in fracture (see Ref. 21 in Chapter 2).

Smith[18] also determined several values of K_I and K_{II} for successive isochromatic fringes and then extrapolated the data to $r = 0$ to obtain the true values of K_I and K_{II}. By using this procedure, the theoretical errors in eq. (2.34) caused by large r values and the experimental errors caused by the crack tip blunting under load and the plastic-yield zone surrounding the crack tip are avoided.

In addition to the references cited above, a detailed example of the use of two-dimensional photoelasticity for solving two-dimensional problems in fracture mechanics is described in Ref. 19. In this reference, the notch-root-radius approach which utilizes the similarity between the stress fields in the vicinity of sharp cracks and notches with finite radius[20] is used to determine the stress-intensity factor. The stress-intensity factor for Mode I type of crack opening can then be represented as

$$K_I = \lim_{\rho \to 0} \sqrt{\frac{\pi}{2}} \sqrt{\rho} \; \sigma_{\max} \qquad (6.3)$$

where ρ = the notch-root radius

          $\sigma_{\max}$ = the maximum stress at the notch root

The experimental procedure for solving a boundary-value problem in fracture mechanics is to conduct a series of photoelastic experiments with decreasing notch-root radii until K_I computed by eq. (6.3) reaches a stationary value. Details of such procedure are described in Ref. 19 where it was found that a minimum notch-root radius of 0.047 in. was sufficient to reach a stationary value of computed K_I. This use of a relatively large notch-root radius circumvents the high-stress concentration and plasticity effect at the crack tip, thus eliminating the major experimental difficulties described previously.

Recently, Marloff et al.[21] suggested two other procedures for determining stress-intensity factors by photoelasticity. One such procedure as modified by this author is to integrate τ_m in eq. (6.1) along a radial coordinate of fixed inclination θ_1 which yields

$$[K_I \sin \theta_1 + 2K_{II} \cos \theta_1]^2 + [K_{II} \sin \theta_1]^2 = \left\{ \sqrt{\frac{2\pi}{r_1}} \int_0^{r_1} \tau_m \, dr \right\}^2$$

$$(6.4)$$

Subsequent numerical procedure for separating out K_I and K_{II} follows that of the direct approach described in Ref. 22. This procedure has the advantage of reducing the random scatter in experimental data by averaging the maximum shear stress across a small line segment of x but requires, by virtue of eq. (2.34), that the elastic-stress distribution be maintained.

The other procedure is to plot τ_m in eq. (6.1) as a function of $r^{-1/2}$ for a given radial inclination of θ_1 which should result in a straight line having a slope of $1/(2\sqrt{2\pi}) \{[K_I \sin \theta_1 + 2K_{II} \cos \theta_1]^2 + [K_{II} \sin \theta_1]^2\}^{1/2}$. Having determined such slope, the subsequent numerical procedure follows that of the direct approach. It should be noted that recent advances in the method of finite-element analysis and the general availability of computer programs for solving two-dimensional problems in elasticity have made two-dimensional photoelasticity somewhat unattractive for solving static-boundary-value problems[23] in fracture mechanics. The decision as to whether one should use computer analysis or photoelastic analysis for such problems hinges on the relative accessibility of each method at the time the solution is sought.

Thermoelasticity Analysis

There are, at this time, two areas in two-dimensional fracture mechanics which cannot be analyzed within reasonable effort or within sufficient accuracy by the method of finite-element analysis. One such area is fracture in the presence of thermal stresses and the other area is fracture dynamics. Due to numerical inaccuracy in the transient-temperature determination, fracture-mechanics problems generated by thermal stresses should be analyzed by other means. One such means is photothermoelasticity where the birefringent response is generated by the self-equilibrated thermal stresses which, in turn, are generated by the nonlinear temperature distribution in the photoelastic model. Both transient and steady-state thermal stress can be determined by this procedure.[24]

The necessary temperature gradient, which simulates that in the prototype, can be applied by either raising or lowering the temperature in the photoelastic model. Since the mechanical and optical properties of commonly used photoelastic materials change rather abruptly at temperatures slightly above room temperature while remaining relatively stable at low temperature, temperature gradients for photothermoelasticity are usually generated by chilling rather than heating the photoelastic model.   Figure 6.1 shows such an arrangement where the stress-intensity factor of a radial crack at the interior wall of a partially filled annulus was investigated.[25]   The temperature gradient was generated by chilling the interior surface of a photoelastic model with a mixture of alcohol and dry ice which was kept at constant temperature by circulating through a large reservoir.   The embedded-polariscope technique shown in Fig. 6.1 provided isochromatics of the central section of the model where the heat flow is within the cross section of the annulus.   A radial crack modeled by a jeweler's sawcut was also introduced in some models. Figure 6.2 shows a typical composite picture of an isochromatic pattern for a steady-state temperature gradient in an uncracked and cracked cylinder.   Typical butterfly-fringe pattern associated with crack tips can be seen in this figure.

Stress-intensity factors were determined by first postulating that the stress in the vicinity of the crack tip is of the $1/\sqrt{r}$ singularity and then by using eq. (6.1) for $\theta = \pm 90$ deg which yields

$$\tau_m = \frac{(K_I^2 + K_{II}^2)^{1/2}}{2\sqrt{2\pi r}} \quad \text{for} \quad \theta = \pm 90 \text{ deg}$$

$$\tau_m = \frac{K_{II}}{\sqrt{2\pi r}} \qquad \qquad \text{for} \quad \theta = 0 \text{ deg} \qquad (6.5)$$

By computing two maximum-shear stresses from the isochromatic fringes closest to the crack tip at perpendicular and parallel distances of r, one can determine the opening and shearing modes of stress-intensity factors, K_I and K_{II}, by the use of eq. (6.1).

Recently, Emery used the same technique to determine stress-intensity factors in edge-cracked plates which are suddenly chilled on the cracked side.[26]   Figure 6.3 shows a composite picture of the isochromatic-fringe patterns for a crack as well as a 90-deg notch in an edge-notched plate.   The photoelastic patterns indicate that a singular state of stress exists at these geometric discontinuities. For the sharp crack, within ± 4 percent of experimental accuracy, the singularity is shown to be of the order of $1/\sqrt{r}$ which is the familiar singularity in fracture mechanics.   The change in stress-intensity factor with time is shown in Fig. 6.4.   The magnitude of the maximum value, as well as the time to reach such maximum value, varies with geometry and surface heat-transfer coefficient.

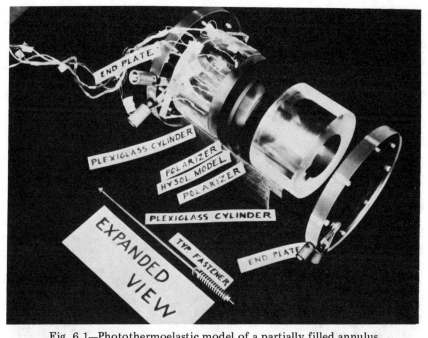

Fig. 6.1—Photothermoelastic model of a partially filled annulus.

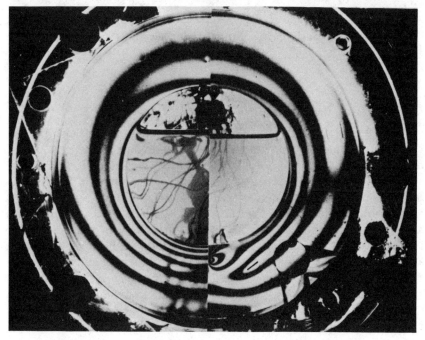

Fig. 6.2—Isochromatic-fringe patterns in a partially filled annulus.

Fig. 6.3—Transient isochromatics for an edge crack (*left*) and a 90-deg notch (*right*) in an edge-cooled plate.

6.3 DYNAMIC PHOTOELASTICITY

Dynamic-photoelasticity Setup

The application of dynamic photoelasticity to transient analysis in fracturing plates was first made by Wells and Post.[11] A Cranz-Schardin type of multiple-spark light source was used to obtain four sequence photographs of transient isochromatic patterns in a fracturing edge-notched specimen of CR-39 plastic. Fringe multiplication and absolute-retardation methods were used to increase sensitivity and to separate the stresses in this experiment. Unfortunately, due to the divergence in loading rates in points throughout the specimen and the strain-rate sensitivity of the material-fringe constant,[27] the state of stress could not be determined accurately.

Recently, Bradley[12] used a modified Cranz-Schardin camera system and a dynamic polariscope, similar to the system used by Dally and Riley,[28] to record the elastic field surrounding a propa-

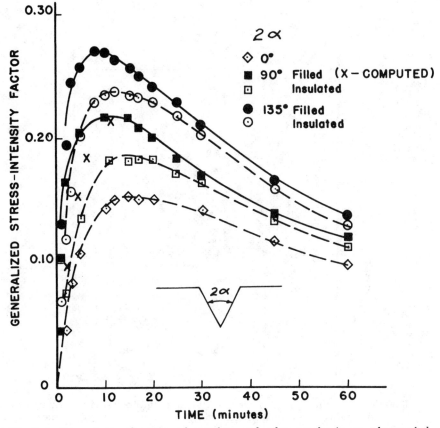

Fig. 6.4—Stress intensification of notches and edge cracks in an edge-cooled plate.

gating crack. The system is composed of a 16 triggerable-spark-gap light source, a 16-in.-diam parallel-light transmission polariscope, and a 16-lens, 11 in. × 14 in. view camera. The discharge of the 0.005-microfarad 30-KV capacitors across the 5-in. spark gaps is initiated by a 70-KV pulse from the trigger transformers which results in a spark duration of 1.25 μsec. The light from each spark gap is picked up by an EG&G Lite-Mike and recorded with Polaroid film on a Tektronix 547 oscilloscope, together with previously recorded timing marks, to provide a permanent record of the firing time of each spark.

The loading fixture shown in Fig. 6.5 is of the fixed-grip design, utilizing a wedging system capable of hand-generated loads up to 2.5 tons. The displacement at the top and bottom edges of the specimen

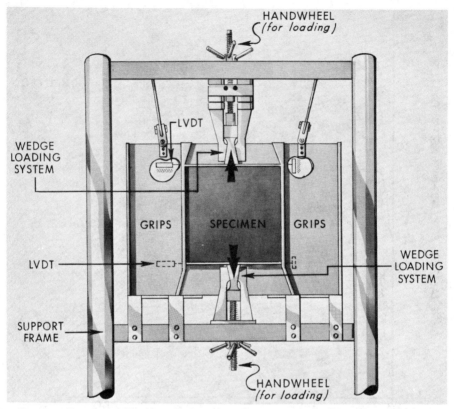

Fig. 6.5—Fixed-grip loading fixture for dynamic photoelasticity. Specimen dimensions: 10-in. effective length × 10-in. width × 3/8-in. thickness.

was measured by two LVDTs with their outputs recorded on an X-Y recorder to allow monitoring of the specimen displacements during the loading cycle. The signals from these two LVDTs were also fed to an oscilloscope to record the grip displacements during the dynamic-fracturing portion of the test. The choice of this loading fixture provides versatility in specimen loading in that, besides uniform strain fields, increasing or decreasing strain fields along the width of the specimen may be selected. Photoelastic models of a single-edge cracked specimen loaded in a fixed-grip configuration were fractured, and the dynamic isochromatics shown in Fig. 6.6 and Fig. 6.7 were recorded.

The maximum velocity of crack propagation was found to be about 15,300 in./sec for single cracks and branching cracks for tests conducted with 0.375-in.-thick Homolite-100 specimens. These results also agree with the maximum velocity of 15,000 in./sec reported

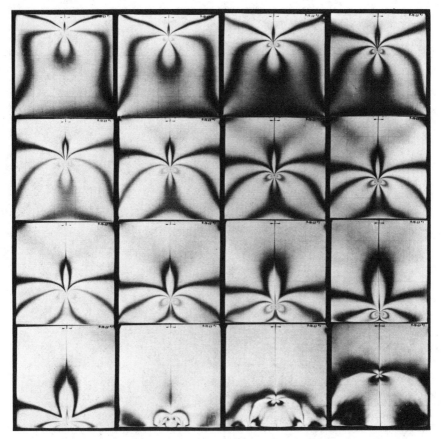

Fig. 6.6—Isochromatic patterns of a constant-velocity crack.

by Beebe[29] for 26.5-in.-long by 13.5-in.-wide tension specimens of Homolite-100.

Dynamic-stress-intensity Factor

Once the isochromatic patterns characterizing the elastic fields associated with the dynamic crack are obtained, it is necessary to convert the isochromatics into quantities which characterize fracture dynamics of brittle materials. The stress-intensity factor, K_I, provides a convenient representation of the dynamic elastic fields associated with the propagating-crack tip and, therefore, several methods were attempted for determining the dynamic-stress-intensity factor from the isochromatic patterns. The direct approach described previously was modified by Irwin[11] for the calculation of K_I

Fig. 6.7—Isochromatic pattern of a branching crack.

from isochromatic patterns obtained by Wells and Post. The maximum shear stress in terms of polar coordinates (r,θ) with the origin at the tip of the crack was represented by

$$(2\tau_{max})^2 = \left\{\frac{K_I}{\sqrt{2\pi r}}\sin\theta + \sigma_{ox}\sin\frac{3\theta}{2}\right\}^2 + \left\{\sigma_{ox}\cos\frac{3\theta}{2}\right\}^2 \quad (6.6)$$

where σ_{ox} is the unknown pseudoboundary stress at a given distance of $x = c$. Irwin also pointed out that a second equation could be obtained by noticing that the $\partial\tau_m/\partial\theta = 0$ at the tip of each isochromatic loop. With these two equations, the unknown quantities of K_I and σ_{ox} can be determined.

Applying this method to rapidly extending cracks in CR-39 from tests conducted by Wells and Post, Irwin obtained results which indicated a general correlation between the stress-intensity factor and the crack velocity in the initial acceleration phase of propagation, but found that the stress-intensity factor continued to increase while the velocity began to level off to a constant value. This method was again used by Pratt and Stock[9] and by Stock,[30] where it was found that the procedure was applicable to r_m[§] values as large as one-half the length of the crack. Stock pointed out that the method produced increasingly better results with the application of smaller and smaller values of r_m and developed an asymptotic technique of evaluating K_I by utilizing information from several isochromatic loops. Wells and Post pointed out in their closure to Irwin's discussion[11] that the method was sensitive to the value of θ_m.[§] Using ±2 deg as a reasonable estimate of the accuracy of the location of the tip, they found that a 9-percent uncertainty in the value of the stress-intensity factor was produced for a nominal value of $\theta_m = 80$ deg. In fact, the sensitivity of the method to the angular value increases with decrease in the magnitude of θ_m and approaches infinity at a value of 69.4 deg with the uncertainty in K_I varying from 3 percent at 90 deg to 70 percent at 72 deg.

To reduce the sensitivity of the above method to θ_m, Bradley modified the above procedure by assuming

$$\sigma_{ox} = \sigma_\infty \quad \text{or} \quad \sigma_{ox} = \frac{K_I}{\sqrt{\pi a}} \tag{6.7}$$

and, thereby, reducing Irwin's method to a one-parameter characterization. Then, by substituting eq. (6.7) into (6.6), he obtained

$$\tau_{max} = \frac{K_I}{2\sqrt{2\pi r}} f(\theta, r, a)$$

where

$$f(\theta, r, a) = \left\{ \sin^2 \theta + 2\sqrt{\frac{2r}{a}} \sin \theta \sin \frac{3\theta}{2} + \frac{2r}{a} \right\}^{1/2}$$

Selecting a node on each of two isochromatic loops at (r_1, θ) and (r_2, θ), the corresponding maximum shear stresses τ_1 and τ_2 can now be subtracted to form

$$\tau_2 - \tau_1 = \frac{K_I}{\sqrt{2\pi}} \frac{f_2 \sqrt{r_1} - f_1 \sqrt{r_2}}{\sqrt{r_1 r_2}}$$

or solving for K_I

[§] r_m and θ_m represent, in terms of the polar coordinate, the location of the tip of the isochromatic where $\partial \tau_m / \partial \theta = 0$.

$$K_I = \frac{2\sqrt{2\pi}\,(\tau_2 - \tau_1)\sqrt{r_1\,r_2}}{f_2\sqrt{r_1} - f_1\sqrt{r_2}} \qquad (6.8)$$

This technique is considerably less sensitive to θ_m showing an uncertainty in K_I of less than 1 percent for 2-deg change. In addition, it does not exhibit the large increase in sensitivity with the reduction in the angular value. A comparison of the K_I values calculated by this method with the more precise method showed good agreement for the use shown in Fig. 6.8 where the values of the stress-intensity factor obtained from both methods were plotted as a function of time. Also plotted in this figure is the crack velocity which exhibits a lag in changes with respect to the changes in the dynamic-stress-intensity factor.

Whereas this method yields satisfactory estimations of the stress-intensity factor for dynamic fracture, a plot of the isochromatics calculated by eq. (6.6) using these values of K_I and σ_{ox} is low by

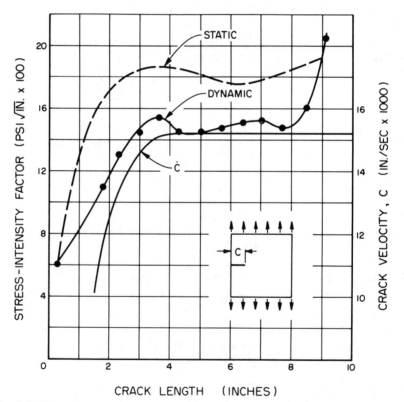

Fig. 6.8—Dynamic stress-intensity factor and crack velocity as a function of crack length calculated by the modified Irwin technique for iso-chromatics shown in Fig. 6.6.

approximately 6 to 8 percent. This error is caused by the prediction of σ_{ox} which is probably too low with respect to its time value. Although the effect of σ_{ox} on the calculation of τ_m is significant, its effect on the calculation of the stress-intensity factor is considerably diminished due to the partial cancelling of this error by the averaging of the two points in eq. (6.8).

6.4 THREE-DIMENSIONAL PHOTOELASTICITY

Since three-dimensional programs in finite-element analysis are not readily available and are also very expensive to run, three-dimensional photoelasticity emerges as a competitive tool for analyzing three-dimensional problems in fracture mechanics. The two popular experimental procedures in three-dimensional photoelasticity are the scattered-light method and the frozen-stress method, but only the latter procedure has been used to solve problems in fracture mechanics.

Scattered-light photoelasticity enables one to analyze three-dimensional problems under live load, but the isochromatic fringes can only be interpreted if the amount of rotation of the refraction tensor is known.[19] In general, this amount of rotation is unknown and, therefore, scattered-light photoelasticity is, in practice, limited to the determination of stresses in the plane of symmetry where such rotation does not occur. Frozen-stress method, on the other hand, requires care against excessive crack blunting when the photoelastic model of the cracked structure is cured at elevated temperature. In the following, an example is taken from Ref. 21 where stress-intensity factors are determined by frozen-stress three-dimensional photoelasticity.

Figure 6.9 shows an epoxy photoelastic model of a turbine disk with tangential cracks emanating from through-bolt holes. Epoxy weights were also cemented to the outside rim of the disk to simulate the centrifugal forces due to turbine blades. The model was then rotated at a speed of 602 rpm and placed through the regular stress-freezing cycle. Figure 6.10 shows the isochromatic-fringe patterns of a 1/4-inch center slice of the turbine disk. Figure 6.11 shows the normalized stress-intensity factors determined by (I) notch-root-radius approach of eq. (6.3); (II) average-stress approach of eq. (6.4); and (III) the slope approach. The normalizing stress σ_N was taken as the hoop stress in an equivalent solid disk. These experimental results are compared with Bowie's solution for an infinite plate with single and double radial crack emanating from a hole and subjected to uniform stress at infinity.[31] Also shown in Fig. 6.11 is Bueckner

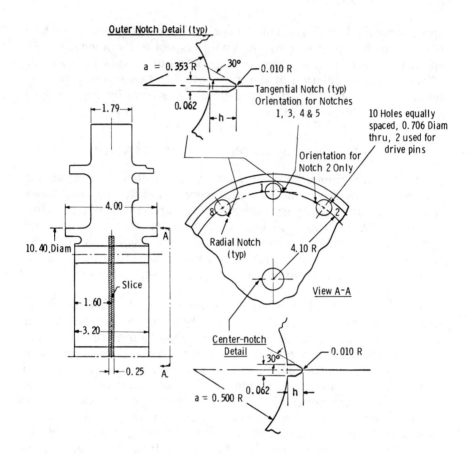

Outer Notch Detail (typ)

a = 0.353 R   30°   0.010 R

0.062   Tangential Notch (typ)
Orientation for Notches
1, 3, 4 & 5

10 Holes equally
spaced, 0.706 Diam
thru, 2 used for
drive pins

Orientation for
Notch 2 Only

1.79

4.00

10.40 Diam

Slice

1.60

3.20

0.25   A.

Radial Notch
(typ)

4.10 R

View A-A

Center-notch
Detail   30°   0.010 R

a = 0.500 R   0.062   h

Notch Specifications

Notch	Notch Direction	h	h/a	N(rpm)	σ_n^*
1	Tangential	0.164	0.46	602	14.7*
2	"	0.162	0.46	"	"
3	"	0.101	0.29	"	"
4	"	0.046	0.13	"	"
5	"	0.164	0.46	"	"
6	Radial	0.099	0.28	"	"
7	"	0.043	0.12	"	"
8	"	0.164	0.46	"	"
Center	—	0.068	0.14	"	18.9*

* Calculated - Disk Computer Program

Fig. 6.9—Photoelastic model of a gas-turbine disk singly notched at center and
through bolt holes. (From Ref. 21)

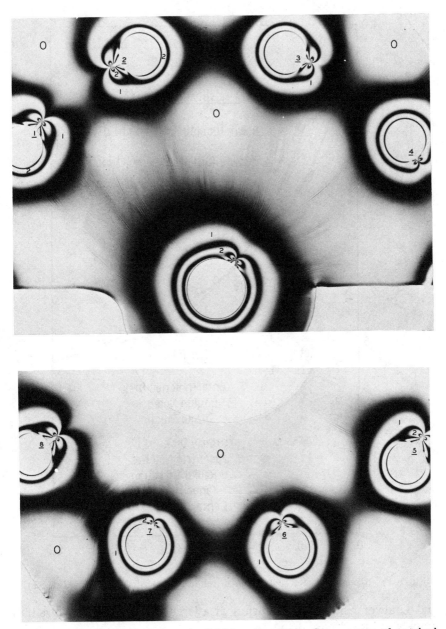

Fig. 6.10—Isochromatic pattern in 0.25-in. slice taken from center of notched
turbine disk; notch number corresponds to those in Fig. 6.9. (From
Ref. 21)

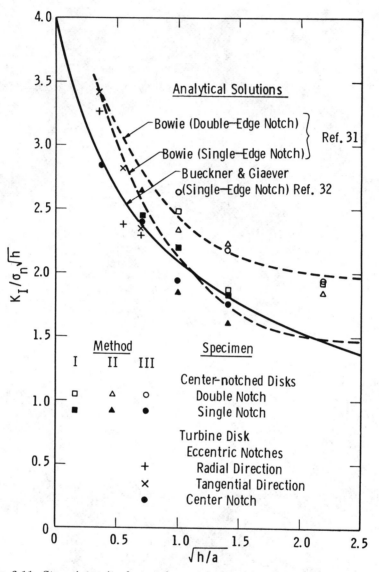

Fig. 6.11—Stress-intensity factors for notched turbine disks. (From Ref. 21)

and Graever's numerical solution of a large disk with a single radial crack at a hole.[32]

6.5 REMARKS

In the preceding section, the use of photoelasticity for determining stress-intensity factors in several problems in fracture mechanics was discussed. In particular, emphasis was placed on experimental procedures, and evaluation techniques peculiar to problems in fracture mechanics were described on the assumption that the readers are familiar with the basic experimental techniques in photoelasticity.

Fracture-mechanics applications of birefringent-coating analysis were omitted due to the questionable thickness effect which becomes pronounced in the vicinity of the crack tip.[19] Also, the application of photoplasticity to problems in fracture was not discussed, since the basic theories in two- and three-dimensional photoplasticity are not well established at this time. Three-dimensional photoplasticity is one area where optical techniques can make substantial contribution to ductile-fracture theory in the future and, thus, warrants further investigation.

REFERENCES

1. Dixon, J. R., "Stress Distribution around a Central Crack in a Plate Loaded in Tension: Effect of Finite Width of Plate," *Jnl. of Roy. Aero. Soc.*, 64, 141-145 (Mar. 1960).
2. Dixon, J. R., "Stress Distribution around Edge Slits in a Plate Loaded in Tension: Effect of Finite Width of Plate," *Jnl. of Roy. Aero. Soc.*, 66, 320-326 (Dec. 1961).
3. Dixon, J. R., and Strannigan, J. S., "Effect of Plastic Deformation on the Strain Distribution around Cracks in Sheet Materials," *Jnl. of Mech. Engrg. Sci.*, 6 (2), 132-136 (1964).
4. Dixon, J. R., "Stress and Strain Distributions around Cracks in Sheet Materials Having Various Work-Hardening Characteristics," *Intl. Jnl. of Fract. Mech.*, 1 (3), 224-244 (1965).
5. Dixon, J. R., "Effects of Crack-Front Geometry and Plate Thickness on the Stress Distribution in Cracked Plates," *Physical Basis of Yield and Fracture—Conf. Proc., Roy. Soc.*, 285 , Series A (1400), (1965).
6. Gerberich, W., "Stress Distribution about a Slowly Growing Crack Determined by the Photoelastic Coating Technique," *Exp. Mech.*, 2 (12), 359-365 (1962).
7. Hubbard, R. P., "Crack Growth under Cyclic Compression," *Jnl. of Basic Engrg.*, Trans. of ASME, 91, Series D (Dec. 1969).
8. Van Elst, H. C., "The Intermittent Propagation of Brittle Fracture," *Trans. of Met. Soc. of AIME*, 23, 460-469 (1964).
9. Pratt, P. L., and Stock, T. A. C., "The Distribution of Strain about a Running Crack," *Conf. Proc., Roy. Soc.*, 285, Series A, 73-82 (1965).
10. Dixon, J. R., and Strannigan, J. S., "Stress Distribution and Buckling in Thin Sheets with Central Slits," *Proc. of the 2nd Intl. Conf. on Fracture*, Brighton (Apr. 1969).
11. Wells, A. A., and Post, D., "The Dynamic Stress Distribution Surrounding a Running Crack—A Photoelastic Analysis," *Proc. of SESA*, 16 (1), 69-96 (1958).

12. Bradley, W. B., and Kobayashi, A. S., "An Investigation of Propagating Cracks by Dynamic Photoelasticity," *Exp. Mech.*, 10 (3), 106-113 (Mar. 1970).

13. Leven, M. M., "Stress Distribution in the M4 Biaxial Fracture Specimen," Westinghouse Res. Rept. 65-1D7-STRSS-RI (Mar. 15, 1965).

14. Leven, M. M., "Stress Distribution in a Pressurized Cylinder with External Longitudinal Notches," Westinghouse Res. Rept. WERL-111402 (Aug. 1965).

15. Post, D., "Isochromatic Fringe Sharpening and Fringe Multiplication in Photoelasticity," *Proc. of SESA*, 12 (2), 143-156 (1958).

16. Dally, J. W., and Ahimaz, F. J., "Photographic Method to Sharpen and Double Isochromatic Fringes," *Exp. Mech.*, 2 (6), 170-175 (1965).

17. Jessop, H. T., "On the Tardy and Senarmont Methods of Measuring Fractional Relative Retardation," *Brit. Jnl. Appl. Phys.*, 4, 138-141 (1953).

18. Smith, D. G., and Smith, C. W., "Photoelastic Determination of Mixed Mode Stress Intensity Factors," *Intl. Jnl. of Engrg. Fract. Mech.*, 4, 357-366 (June 1972).

19. Kobayashi, A. S., "Photoelastic Studies of Fracture," *Handbook of Fracture*, 3, 311-369, Academic Press (1971).

20. Creager, M., "The Elastic Stress Field Near the Tip of a Blunt Crack," Lehigh University, Institute of Res. Rept. (Oct. 1966).

21. Marloff, R. H., Leven, M. M., Ringler, T. N., and Johnson, R. L., "Photoelastic Determination of Stress Intensity Factors," *Exp. Mech.*, 11 (12), 529-539 (Dec. 1971).

22. Dally, J. W., and Riley, W. F., "Separation Techniques," *Experimental Stress Analysis*, McGraw-Hill, 235-247 (1965).

23. Drucker, D. C., "Thoughts on the Present and Future Interrelation of Theoretical and Experimental Mechanics," *Exp. Mech.*, 8 (3), 97-106 (Mar. 1968).

24. Becker, H., "An Exploratory Study of Stress Concentrations in Thermal Shock Fields," *Jnl. of Engrg. Ind., Trans. of ASME* (1962).

25. Emery, A. F., Barrett, C. F., and Kobayashi, A. S., "Temperature Distributions and Thermal Stresses in a Partially Filled Annulus," *Exp. Mech.*, 6 (12), 602-608 (Dec. 1966).

26. Emery, A. F., Williams, J. A., and Avery, J., "Thermal-stress Concentration Caused by Structural Discontinuities," *Exp. Mech.*, 9 (12), 558-564 (Dec. 1969).

27. Clark, A. B. J., and Sanford, R. J., "A Comparison of Static and Dynamic Properties of Photoelastic Materials," *Exp. Mech.*, 3 (6), 148-151 (June 1963).

28. Dally, J. W., and Riley, W. F., "Stress Wave Propagation in a Half-plane Due to Transient Point Loading," *Developments in Theoretical and Applied Mechanics*, 3, 357-377, Pergamon Press (1967).

29. Beebe, W. M., "An Experimental Investigation of Dynamic Crack Propagation in Plastics and Metals," Cal. Tech. Rept. No. AFML-TR-66-249 (Nov. 1966).

30. Stock, T. A. C., "Stress Intensity Factors for Propagating Brittle Cracks," *Intl. Jnl. of Fracture*, 3 (2), 121–130 (June 1967).
31. Bowie, O. L., "Analysis of an Infinite Plate Containing Radial Cracks Originating at the Boundary of an Internal Circular Hole," *Jnl. of Math. and Physics*, 35, 60–71 (Apr. 1956).
32. Bueckner, H. F., and Giaever, I., "The Stress Concentration in a Notched Rotor Subjected to Centrifugal Forces," *Zut. Angeew. Mathematika and Mechanik*, 46 (5), 265 (1966).

AUTHOR INDEX

SUBJECT INDEX